Genetic Basis of Biochemical Mechanisms of Plant Disease

Edited by
James V. Groth and William R. Bushnell

APS PRESS

The American Phytopathological Society
St. Paul, Minnesota

Cover: Haustorium of *Puccinia coronata* and associated host organelles within host cell (for full description, see page 68). Reprinted, by permission of Academic Press, from "Structure and Physiology of Haustoria" by D. E. Harder and J. Chong, pages 431–476 in *The Cereal Rusts*, Volume I, edited by William R. Bushnell and Alan P. Roelfs. Copyright © 1984 by Academic Press, Inc.

This publication is based on presentations from a symposium entitled "Genetic Basis of Biochemical Mechanisms of Disease" held in conjunction with the annual meeting of The American Phytopathological Society, August 13, 1984, in Guelph, Ontario. To make the information available in a timely and economical fashion, this book has been reproduced directly from typewritten copy submitted in final form to The American Phytopathological Society by the editors of the volume. No editing or proofreading has been done by the Society.

Library of Congress Catalog Card Number: 85-073193
International Standard Book Number: 0-89054-068-3

Second Printing, 1986

Printed in the United States of America

The American Phytopathological Society
3340 Pilot Knob Road
St. Paul, Minnesota 55121, USA

CONTENTS

PREFACE

Interest in biochemical mechanisms of host-pathogen interaction has been led by two distinct groups of researchers: one stressing biochemistry-physiology and the other stressing genetics. The two groups have tended to attack problems with different tools and philosophies. The result has often been that they either fail to understand one another or that they have not designed experiments as efficiently as possible. It appears that this is changing. First and foremost, the rapid and massive appearance of new technologies in biology has forced researchers who use them to broaden their perspectives; answers obtained are of direct interest to geneticists and biochemists alike. Second, more people have realized that narrow, insular approaches to the complex questions are less likely to provide realistic answers. The present symposium, held August 13, 1984, at the APS-CPS meeting at Guelph, Ontario, reflects this trend toward a more interdisciplinary approach to host-parasite interaction. The contributions span coevolution, pathogenesis, and cytology as well as genetics and biochemistry-physiology. As the title suggests, each contribution includes aspects of the genetics of plant disease. The biochemical mechanisms of plant disease, as yet to be elucidated in a definitive way, are also examined using empirical and logical arguments.

The five chapters in this symposium are distinct, both in subject matter and in style. Where the same issues are addressed in more than one chapter, they are likely to be addressed differently. Yet there is a great deal of complementation among contributions. Perhaps the most significant impression that can be obtained from a comparison of contributions is that there is not likely to emerge one explanation of host-parasite specificity as the breakthroughs occur. In fact, a demonstration of the complexity of the problem was one of the primary objectives of the symposium. A universal model of the mechanism of host-parasite interactions will have to be painted with a broad brush indeed if it is to be truly universal. Details vary from disease system to disease system, or even from one host-line:pathogen-isolate combination to another within the same disease system. There is not even clear agreement of what constitutes a "classical" gene-for-gene interaction. As happens whenever a relatively simple hypothesis is further examined, many variations on the idealized gene-for-gene pattern are now documented. This makes a precise, logical definition of gene-for-gene relations all the more imperative, but perhaps also all the more elusive. Among other possible criteria for gene-for-gene relations, they can be characterized genetically as possessing primarily nonadditive gene interactions, that is interactions among different host resistance genes and among different pathogen virulence genes which are strongly dominant and epistatic to the extent that one allele totally masks the effect of all others at the same locus or different loci. Unless one presupposes that there is something mechanistically unique about one-gene for one-gene interactions, there is really no reason to think that they are qualitatively distinct from, say, one-gene for two-gene interactions, or from other interactions not involving primarily additive gene effects. The repeated finding that subtle differences in compatibility can be highly pathogen "race" or host "line" specific would seem to extend the importance of gene-for-gene interactions if defined very broadly. It also leaves more room for the inclusion of a higher proportion of additive gene action, which has also been demonstrated in a few systems. These are primarily genetic arguments. The symposium contains information which extends the complexity of classifying host-parasite interaction using arguments which are based on biochemistry, pathogenesis, and coevolution, to name three other areas of investigation.

The symposium on which this book is based was organized chiefly by W. R. Bushnell of the Disease and Pathogen Physiology Committee of APS. The editing of this book was done chiefly by J. V. Groth of the Genetics Committee of APS. No attempt was made in the editing process to alter content except for the sake of clarity, although we were not in agreement with the conclusions in every instance. We thank the authors for their excellent contributions and we also thank Colleen Curran who copyedited and collated all chapters and Brenda Anderson who typed the entire book.

J. V. Groth
Department of Plant Pathology
University of Minnesota
St. Paul, MN 55108

W. R. Bushnell
Cereal Rust Laboratory
U. S. Department of Agriculture
and
Department of Plant Pathology
University of Minnesota
St. Paul, MN 55108

Genetic Basis of Biochemical Mechanisms of Plant Disease

THE ROLE OF NONSELF RECOGNITION IN PLANT DISEASE RESISTANCE[1,2]

W. R. Bushnell

Cereal Rust Laboratory,
Agricultural Research Service, U.S. Department of Agriculture and Department of Plant Pathology,
University of Minnesota, St. Paul, MN 55108.

Several authors have invoked recognition of nonself as the means by which plants recognize that an invading organism is to be rejected. In this view, the plant turns on its various defenses and repels attack because the invader is recognized as nonself or a foreign entity. Thus the plant somehow distinguishes between self and nonself. Callow (36) stated that "...the potential host plant may be able to detect or recognize a potential fungal pathogen as foreign or 'non-self' and use the initial act of recognition to trigger a range of individual resistance mechanisms." The compatible pathogen is able "...to suppress or divert...general or non-specific recognition so that the fungus is recognized

[1] Cooperative investigations, Agricultural Research Service, U.S. Department of Agriculture and Department of Plant Pathology, University of Minnesota. Paper No. 14250, Misc. Journal Series, Minnesota Agricultural Experiment Station.

as 'self', resulting in disease." Other authors have indicated that plants have an ability to recognize "nonself" or "foreign entities", implying that this ability is an important component of plant disease resistance (170,187,208). Still other authors have pointed out that an ability for nonself recognition is an hypothesis worth investigating with respect to plant disease (128,229). Whether nonself (or self) is meant in a direct, literal sense by these authors is not clear. In any case, none have critically examined the evidence for and against the possibility that plants have a generalized ability to distinguish between self and nonself. It is the purpose of this chapter to make such an examination.

An ability to recognize nonself can be either based on a direct active recognition of nonself, or be an indirect passive result of active recognition of self. Which of these is operating depends on the biological system involved and is not always known (135). Except as explicitly indicated, distinction between these alternatives is not intended when "nonself recognition" is used in the discussion that follows.

EVIDENCE FOR NONSELF RECOGNITION IN PLANTS

Responses to Pathogens

The idea that plants resist potential pathogens because the pathogens are recognized as nonself comes in part from the remarkable ability of plants to reject most attacking microorganisms. The plant is a nonhost for all but the few more or less specially adapted pathogens to which it is susceptible. We can speculate that this broad resistance might be based on a general ability of plants to discriminate between itself and foreign organisms. However, defense responses are more probably triggered by certain activities and substances of attacking organisms (as will be discussed later).

Specific-Purpose Recognition

Another factor that contributes to the concept of nonself recognition in plant disease is the prevalence of specific-purpose systems for recognition of nonself (or self) in both plants and animals. This includes the ability of individual vertebrate animals to reject cells and tissues of other individual animals of the same species or of other species through the workings of the major histocompatibility complex and the animal's humoral immunological system (65,135). Furthermore, both higher plants and fungi have sexual discrimination systems which distinguish between self and nonself within and between species. Some aspects of these systems will be discussed later. Katz (135), who drew upon examples from lower animals, higher plants, and higher animals, concluded that "...self-recognition appears to be a fundamental biological process concerned with control of many types of developmental and differentiation events." He marshaled evidence that the active event in immunological systems is recognition of self instead of recognition of nonself, an argument reminiscent of the controversy on whether the active event in race-cultivar specificity in plant disease is for compatibility or incompatibility (60,61,82). In any case, the prevalence of systems for recognition of nonself (or self) in diverse biological organisms, in itself, suggests that similar and possibly related systems could be operating in plant disease.

Cross-reactive Antigens

A more direct reason for implicating nonself (or self) recognition in plant disease is the occurrence of antigens in parasites which cross-react immunologically with antigens from their hosts. These cross-reactive antigens (often termed common antigens) have been investigated and reviewed extensively by DeVay and coworkers (48,72,73,74). In several diseases involving bacterial or fungal pathogens, one or more antigens have been found in host and pathogen which cross react. These cross reactions have usually been detected with precipitin reactions in agar-gel double diffusion plates (Fig. 1), although enzyme-linked immunosorbent assays (ELISA procedures) are beginning to be used (6).

Enough positive findings of cross-reactive antigens have been reported to suggest that they are generally present and correleated with plant disease, although there are several reported failures to find them. Sometimes the cross-reactive antigens are found only in compatible host-parasite combinations and not incompatible ones for a given disease, as was the case for black rot of sweet potato (Fig. 1), or they are found in greater amounts with compatibility than incompatibility. More often, the cross-reactions do not relate to specificity within a disease (such as specificity at the race or _forma specialis_ levels,)

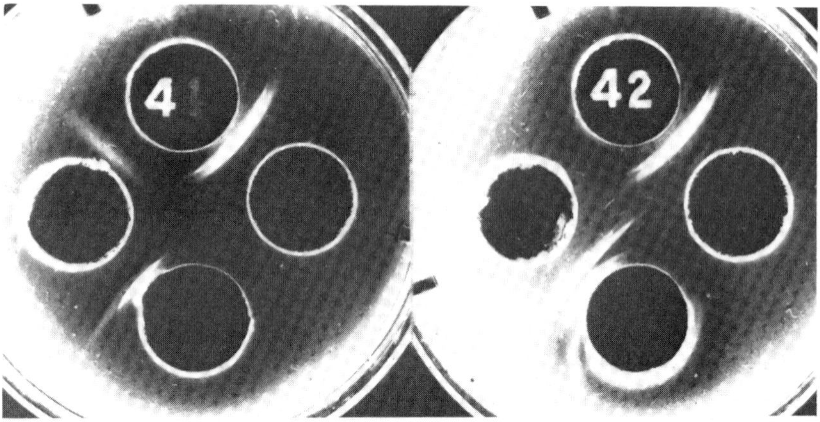

Fig. 1. Example of cross-reactive antigens between host and parasite in agar-gel double diffusion tests. Left dish: top well contained antiserum of _Ceratocystis fimbriata_ isolate 104; left well contained antigen of susceptible sweet potato line UC779. Note precipitin band between the top and left wells. Right dish: similar to left dish except left well contained antigen of resistant sweet potato variety Sunnyside. No precipitin band was produced. Reprinted, by permission, from DeVay _et al_. (73).

but, instead, correlate with the ability of a pathogen species to cause disease in a host species; i.e., they correlate with the presence of basic compatibility between host and parasite. In the evolution of plant disease, selection pressure apparently has favored the presence of certain like antigens in host and parasite.

None of the cross-reactive antigens have been identified and their functions are unknown. Their presence suggests that there is an advantage to a parasite to become like the host with respect to certain antigenic proteins. If the host has an ability to recognize nonself, it follows that cross-reactive antigens might have evolved as a way for the parasite to avoid recognition as nonself.

Some trematode parasites of vertebrates seem to mimic their animal hosts as a way to avoid being recognized as a foreign organism. Some of these parasites are thought to have evolved antigenic surface features that resemble features of the host (Fig. 2), and others cloak themselves with substances from the host (24,63). The importance of these features for pathogen virulence is open to question (24,26,63), but their presence suggests that the disguised parasite has a reduced likelihood of triggering the immunological defenses of its vertebrate host.

Plants do not have immunological defense systems directly comparable to those of vertebrate animals, and we have no defined mechanism to explain why antigenic similarity to a plant host confers an advantage on a parasite. DeVay and Adler (72) speculated that certain parasite antigens, probably proteins, may disturb the metabolism of a potential host and that such disturbance is avoided or lessened when those antigens are made to resemble those of the host. In this view, the parasite is able to avoid provocation by mimicking the host with respect to certain key antigens. This does not imply that the host has a general ability to respond to nonself, but only an ability to respond to certain antigenic features of potential parasites.

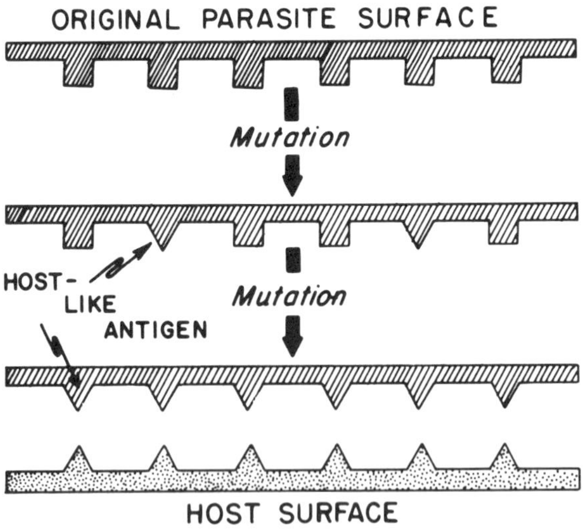

Fig. 2. Evolution of host-like antigens on a parasite's surface as postulated to help the parasite avoid the immune response in animal hosts. Adapted from Damian (63).

Tissue and Cell Incompatibility in Higher Plants

Another phenomenon that suggests that higher plants have an ability to recognize nonself is the fairly common observation that tissues or cells of one species are incompatible with tissues or cells of another species. Examples of this were provided by Teasdale et al. (219) who placed cells and tissues of diverse plants and animals on pea pod endocarp (Table 1) (Figs. 3, 4). All cells and tissues except pea pod endocarp itself and pea pollen were incompatible, generating blackened tissue and triggering phytoalexin production. Although this study was limited in scope, the results support the idea that the pea tissue recognized cells and tissues of all species other than itself. As will be described later, certain pairings between higher plant species in tissue culture or in grafts generate incompatibility reactions. These seem

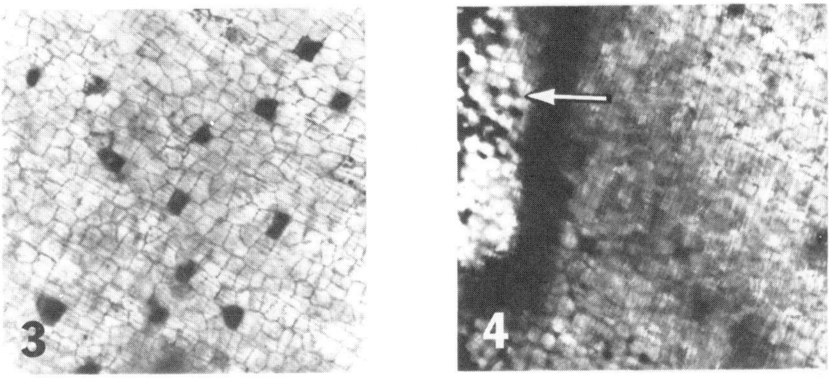

Figs. 3-4. Tissue responses of pea pod endocarp to exposure to bean pollen (Fig. 3) or bean pod tissue (Fig. 4). The darkened areas exhibited yellow-green fluorescence and produced phytoalexins. The bean tissue was in contact with the light region indicated by the arrow. These are representative of incompatiility reactions that can occur between paired organisms as in Table 1. Reprinted, by permission, from Teasdale *et al.* (219).

to be fortuitous cases in which components of one species are injurious to the partner species and are not cases of general nonself recognition.

EVIDENCE AGAINST NONSELF RECOGNITION IN PLANTS

Whether or not plants have a general ability to discriminate between themselves and alien (nonself) organisms can be evaluated by surveying additional cases in which a plant is forced into an intimate association with an alien organism. Such intimate pairings have been constructed with protoplasts, cells, tissues, and organs of various types. Reviewed here will be pairings involving plant protoplasts, tissue cultures and grafts.

Table 1. Compatibility of pea pod endocarp with diverse cells and tissues suggesting that pea (Pisum sativum) can recognize species other than itself as non-self. Data of Teasdale et al. (219).

Tissue or cells	Compatibility[a]
Bean pod endocarp	-
Bean pollen	-
Mouse tumor cells	-
Fusarium solani f. sp. *phaseoli*	-
Fusarium solani f. sp. *pisi*[b]	-
Fusarium nivale	-
Pea pod endocarp	+
Pea pollen	+

[a] + indicates compatibility; - indicates incompatibility

[b] Pathogen of pea

Protoplast Fusion

Isolated protoplasts of plants have been fused with protoplasts and cells of a wide variety of nonself species. For example, there are more than 50 reports of formation of somatic hybrids between species and genera of higher plants by protoplast fusion (66). Many similar hybrids have been produced between species within the animal kingdom (66). Furthermore, interkingdom fusions have been obtained between higher plant species and a wide variety of

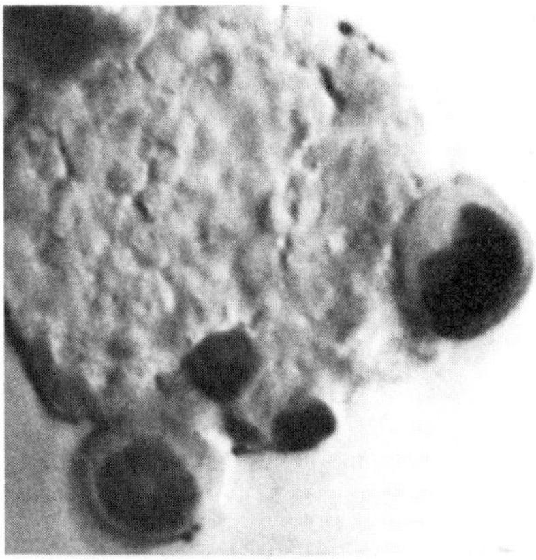

Fig. 5. Fusions between a tobacco protoplast and human HeLa cells, an example of cytoplasmic and nuclear compatibility between divergent organisms. Two HeLa cells have fused with the protoplast, a third is attached to the protoplast surface. The protoplast contains two small tobacco nuclei. The HeLa nuclei retained their integrity for up to six days after fusion. (The nuclei were stained with carbol fuchsin.) Reprinted, by permission, from Jones *et al.* (131). Copyright 1976 by the American Association for the Advancement of Science.

microorganisms and cells or protoplasts of animals (Table 2) (Fig. 5).

In fused protoplasts, the cytoplasm, cytoplasmic organelles, and nuclei of one partner can be found to be in direct intimate contact with cytoplasm of the other partner, although small bodies such as microorganisms and organelles are sometimes engulfed in intact form (98) or encased in vesicles within a recipient protoplast (66). Davey *et al.* (68) showed rigorously by electron microscopy that cytoplasm of the amphibian *Xenopus* mixed intimately with cytoplasm of carrot protoplasts, and similar work has shown that protoplasm of yeasts and algae come into direct contact with recipient higher plant cytoplasm after fusion (67,93).

Fused protoplasts frequently persist for days or weeks without signs of incompatibility between the

Table 2. Diverse protoplasts and cells that have been fused with higher plant protoplasts.

Higher plant protoplast	Fusion partner	References
Vinca	Bacterium spheroplasts	Hasezawa *et al.* (107)
Carrot	Green alga protoplasts	Fowke *et al.* (92,93)
Virginia creeper	Yeast protoplasts and cells	Davey and Power (67)
Mung bean	Amoeba cells	Rajasekhar *et al.* (197)
Belladonna	Amoeba cells	Rajasekhar *et al.* (197)
Carrot	Xenopus cells	Davey *et al.* (68); Ward *et al.* (225)
Tobacco	Avian erythrocytes	Willis *et al.* (227)
Soybean	Drosophila cells	Hadlaczky *et al.* (102)
Tobacco	Hamster cells	Mastrangelo and Mitra (169)
Arabidopsis	Human lymphocytes	Sidorov *et al.* (210)
Carrot	Human Hela cells	Dudits *et al.* (79)
Haplopappus	Human Hela cells	Lima-de-Faria *et al.* (160)
Tobacco	Human Hela cells	Jones *et al.* (131)

components of the two partners. The fused protoplasts usually regenerate walls and sometimes divide, produce callus and differentiate into plants. More often the fused protoplasts do not have enough compatibility to grow as a balanced heterokaryon into a differentiated organism. Nevertheless, they frequently have enough compatibility to allow cytoplasm and nuclei from the fusion partners to coexist for extended times.

Such fusions indicate that "...the mixing of cytoplasm from two higher plants alien to one another, or the introduction of organelles from one species into another does not produce incompatibility reactions..." and that the plasma membranes also must fuse without generating incompatibility (30). If higher plants have a general ability to recognize organisms other than themselves, that ability apparently does not reside within individual protoplasts.

Fungal protoplasts of several species have been used to produce somatic hybrids between fungal species (Table 3), suggesting that fungal protoplasts, like those of higher plants, lack a general ability to recognize species other than themselves. These fungal

Table 3. Fusion of protoplasts between fungal species. Data from Ferenczy (88) and Peberdy (193).

Species A[a]	Species B[a]
Aspergillus nidulans	*Aspergillus fumigatus*
Aspergillus nidulans	*Aspergillus rugulosus*
Aspergillus niger	*Aspersillus rugulosus*
Candida tropicalis	*Saccharomycopsis fibuligera*
Kluyveromyces lactis	*Kluyveromyces fragilis*
Penicillium chrysogenum	*Penicillium cyaneo-fulvum*
Penicillium chrysogenum	*Penicillium roquefortii*
Penicillium citrinum	*Penicillium cyaneo-fulvum*

[a] Protoplasts of species A were fused with protoplasts of species B

hybrids have been formed by pairing complementary auxotrophic mutants and the hybrids definitely have contained genetic information from the parental mutants. However, the hybrids have occurred at much lower frequencies than has been the case with hybrids from higher plant protoplasts, and the events at fusion have not been studied cytologically. Whether fused fungal protoplasts exist as such for extended times has not been demonstrated. Furthermore, the reported fusions have been mostly with closely related species of imperfect fungi which may not be distinct species in all cases. Still, given no evidence to the contrary, it seems likely that fungal protoplasts from different species are compatible with one another.

Fusions between protoplasts of pathogenic fungi and protoplasts of their higher plant host species apparently have not been reported. Since fungal protoplasts are smaller than higher plant protoplasts, they may be engulfed as intact protoplasts as shown by Gerson *et al.* (98) for protoplasts of the yeast-like *Aureobasidium pullulans* in protoplasts of *Zea mays*. Aside from the question of general nonself recognition, such experiments might help determine if recognition mechanisms controlling host-parasite compatibility/incompatibility reside within protoplasts.

Mixtures of strong, impure hydrolytic enzymes have generally been used to isolate both fungal and higher plant protoplasts. These enzymes may alter the surface of plasma membranes and thereby remove an ability for nonself recognition. The results with intact cells in tissue cultures and grafts to be described next suggest that this is not the case.

Tissue Co-culture with Paired Species

Ball (17,18) described a series of experiments in which callus cultures of various different higher plant species were grown in contiguous pairs, either side by side or one on top of the other (Fig. 6), and were examined for signs of incompatibility. Similar studies were conducted by Gautheret (96), Porcelli-Armenise *et al.* (194), and Fujii and Nito (94) with

Fig. 6. Tissue culture of *Lobelia erinus* (white) on which tissue of *Chenopodium album* (green) grew for 10 days. The tissues were compatible with one another, as are many such pairings between species and genera of higher plants. Reprinted, by permission, from Ball (18).

explants from woody species. Typical results are shown in Table 4. Some combinations developed a darkly staining layer which separated the two cultures, indicating an incompatibility barrier reaction, whereas other combinations grew without signs of incompatibility, either without close mutual contact or with an intimate junction between the paired cultures. Ball (18) used 11 species to make 45 pairings between species, only a few of which were incompatible. Fujii and Nito (94) used 6 species of fruit trees and found 8 of 15 pairings between species to be incompatible, whereas the members of the 7 other pairings could be grown as callus on top of each other. Most of the incompatible pairs involved one species, *Vitis vinifera*.

Table 4. Compatibility of higher plant callus cultures grown in contiguous pairs. Data from Ball (17) and Gautheret (96).

Species A[a]	Species B[a]	Compatibility[b]
Acer sp.	*Sambucus sp.*	-
Nicotiana sp.	*Dahlia sp.*	-
Zinnia sp.	*Lobelia erinus*	-
Chenopodium alba	*Amaranthus sp.*	+
Nicotiana tabacum	*Daucus carota*	+
Nicotiana sp.	*Lobelia sp.*	+
Nicotiana sp.	*Dianthus sp.*	+
Populus sp.	*Salix sp.*	+

[a] Callus of species A was grown in contact with callus of species B.

[b] + indicates tissues were compatible; - indicates tissues were incompatible.

Clearly some species are incompatible with certain other species in tissue co-culture, but many, perhaps the majority, of pairings are compatible. These results argue against the idea that higher plant tissues have a general ability to reject alien species based on an ability to recognize nonself.

The paired tissue cultures can influence each other short of barrier reactions in that morphological characteristics or isozyme composition may be altered (17). In this way, one species may show that it recognizes the presence of the other, as a result of molecules moving from one species to the other. However, such morphological responses do not indicate a general ability to recognize nonself.

Grafts

The formation of a fully compatible graft between stock and scion of two higher plants requires not only the absence of detrimental rejection reactions, but also the positive differentiation of functional vascular connections across the graft union. The graft is initiated by growth of undifferentiated callus from stock and scion which adhere and grow into each other within a few days (Fig. 7). Later, vascular tissue differentiates within the callus. The initial callus growth is often produced regardless of the degree of compatibility between stock and scion, at least in the Solanaceae (232). However, incompatibility is sometimes expressed by rapid development of necrosis at the graft union, as shown in Fig. 9 for a Sedum-Solanum graft, a response which at least superficially resembles hypersensitive reactions in plant disease. There is usually no necrotic response when incompatibility is based on inadequate differentiation of vascular tissue.

The physiological reasons for graft incompatibility are not well known, except that pear or peach scions are thought to supply cyanogenic glycosides, which are hydrolysed in quince or almond rootstocks, releasing toxic cyanide (232). Yoeman (231) speculated that incompatibility is triggered by specific recognition events, but little is known about them.

Numerous compatible grafts have been reported between species of higher plants, suggesting that graft incompatibility is not triggered by an ability of one species to recognize another as nonself. For example, Yeoman et al. (232) reported that only 2 of 17 pairings involving 7 genera of the Solanaceae were incompatible (Table 5). Although members of the Solanaceae apparently are more compatible with each other than are species within many other families of higher plants (106), numerous compatible grafts between other species and genera of higher plants have

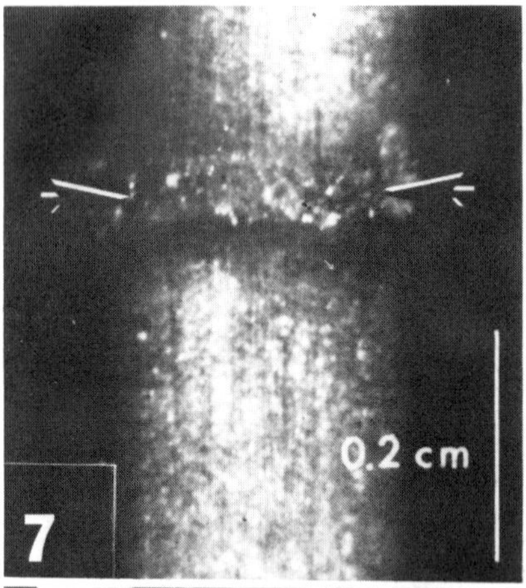

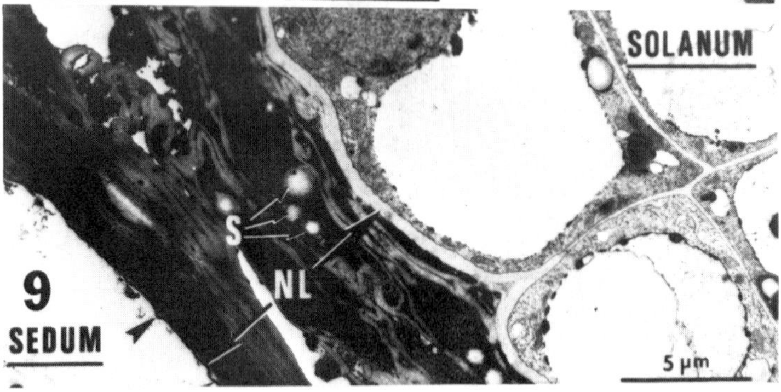

Fig. 7. Compatible homograft with Sedum, showing proliferation and union of callus cells (arrows) from both stock (below) and scion (above). Three days after graft was made. Reprinted, by permission, from Moore and Walker (177).

Fig. 8. Compatible root graft between two trees of <u>Quercus rubra</u>. Vascular tissues functionally interconnect the two trees. Specimen kindly provided by Robert Blanchette.

Fig. 9. Necrotic layer (NL) of collapsed tissue at incompatible graft at Sedum-Solanum interface, 8 days after the graft was made, illustrating a type of graft incompatibility between species. S = starch deposit. Reprinted, by permission, from Moore and Walker (178).

Table 5. Intergeneric grafts within the Solanaceae showing that most pairings are compatible. Data of Yeoman et al. (232).

Stock	Scion: *L. esculentum*	*D. stramonium*	*N. physaloides*	*N. tabacum*	*S. melongena*	*P. hybrida*	*C. frutescens*
	Compatibility[a]						
Lycopersicon esculentum	+	+	–	+	+	+	–
Datura stramonium	+	+	+	+	+	+	+
Nicandra physaloides	–	+	+	+	+	+	
Nicotiana tabacum	+	+	+	+	+	+	
Salanum melongena	+	+	+	+	+	+	
Petunia hybrida	+	+	+	+	+	+	
Capsicum frutescens	–	+					+

[a] + indicates stock and scion were fully compatible; – indicates incompatibility.

Table 6. Natural interspecific and intergeneric root grafts in trees as listed by Graham and Bormann (101).

Species A[a]	Species B[a]
Interspecific	
Acer negundo	*A. platanoides*
Nothofagus solandri	*N. troncata*
Quercus alba	*Q. rubra*
Quercus marilandica	*Q. velutina*
Intergeneric	
Carya sp.	*Quercus nigra*
Carya sp.	*Quercus stellata*
Carya sp.	*Ulmus sp.*
Eugenia jambolana	*Santalum album*
Pinus sylvestris	*Picea abies*
Pinus sylvestris	*Picea banksiana*
Pinus sylvestris	*Picea nigra*

[a] Compatible graft has been reported between Species A and Species B

been reported (106) including natural grafts formed between roots of different tree species and genera (Fig. 8, Table 6).

The chances for formation of a fully compatible graft diminish as the taxonomic distance between the graft partners is increased. Intergeneric grafts are less likely to succeed than interspecific grafts, and interfamily grafts are even less successful (106). This indicates that genetic and physiological similarities between graft partners improve the chances for success in the complex task of forming functional vascular connections and avoiding toxic effects of one partner on the other. While genetic and physiological similarities no doubt favor graft compatibility, and certain pairings of species are incompatible (much as in tissue co-culture as described earlier), graft incompatibility does not appear to result from nonself recognition.

SPECIFIC-PURPOSE NONSELF RECOGNITION SYSTEMS

In contrast to the questionable possibility of a general ability to recognize nonself, plants have several well-known, highly specific systems for distinguishing self from nonself for purposes of protecting or enhancing the genetic composition of species or individuals. Higher plants have systems which regulate compatibility between pistil and pollen; fungi have systems which control sexual compatibility among mating types and control somatic compatibility among vegetative strains. These systems are under strict genetic control, usually involving a small number of genetic loci. Depending on the system involved, genes at these loci must match or mismatch in paired individuals to determine whether members of the pair are compatible or incompatible. Each recognition system fulfills a specific need for the organisms involved. Neither singly nor collectively do these systems function as general systems for recognition of nonself. To illustrate these points, somatic incompatibility systems in fungi and from pistil-pollen systems in higher plants may be used.

Somatic Incompatibility in Fungi

The unprotected protoplasm within the plasmodium of a slime mold is at risk by the invasion of another individual plasmodium of the same species, thereby mixing the cytoplasm and nuclei of the two plasmodial individuals. Slime molds have evolved somatic incompatibility systems that can either prevent fusion between two strains or, after fusion, lead either to the death of one strain (Figs. 10-11) or to selective death of nuclei of one of the strains (87,157,207). This results in protection of the genetic integrity of ecological or geographic strains. Each species studied (mainly Physarum polycephalum and Didymium

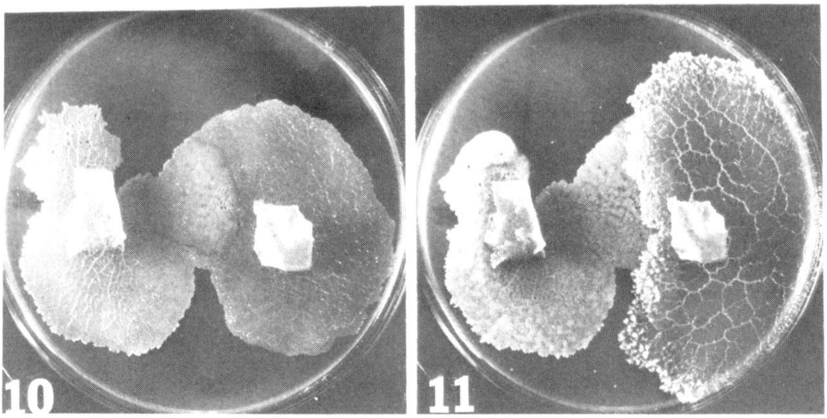

Fig. 10-11. Somatic incompatibility between two strains of *Physarum polycephalum* illustrating a specific-purpose type of nonself recognition which here serves to preserve genetic difference between the strains. Fig. 10, Five and one-half hours after fusion of plasmodia. A dead area extends most of the way between the inoculation blocks; Fig. 11, twenty-three hours after fusion. The sensitive plasmodium on the left is dead, the killer plasmodium on the right is beginning to overgrow the dead area. Reprinted, by permission, from Carlile (37).

iridis) has several genetic loci that condition somatic incompatibility (58,157). If the same gene is not present at each locus in each of the two strains, they will be incompatible. (This is termed heterogenic incompatibility). Genes at some loci condition fusion of plasmodia; genes at other loci condition compatibility after fusion. Similar heterogenic somatic incompatibility systems operate in some hyphal fungi in which hyphal anastomosis allows nuclei and cytoplasm of two strains to intermix, as in *Neurospora crassa*, *Podospora anserina* and also in many wood rotting basidiomycetes (87). As indicated by Bushnell (30), natural cytoplasmic incompatibility in fungi as represented by heterogenic somatic incompatibility by isolates within species, contrasts with the many cases of cytoplasmic compatibility demonstrated by fusion of isolated protoplasts between species. (See earlier section on protoplast fusion). Since neither plasmodia (58) nor hyphae (87) of different species fuse, it follows that somatic incompatibility evolved partly for the purpose of conserving genetic differences (both nuclear and cytoplasmic) among isolates within species, possibly to promote the evolution of new species (87).

Pistil-pollen Compatibility/Incompatibility

Incompatibility between pistil and pollen within a species of flowering plants is genetically controlled mainly by the S-allele system which prevents pollen of an individual plant from being accepted by the pistil of that same plant (31,118). Rejection of pollen occurs when S-alleles in pollen match S-alleles in the pistil. There are numerous alleles at the main S- locus among the interbreeding population of a species so that most individual plants in that population are different from most other individuals. Consequently the pollen of one plant is almost always compatible with the pistil of another plant. By limiting self-fertilization, the S-allele incompatibility system forces outbreeding among individual plants within a species and promotes genetic heterogeneity. Though not well understood, the physiological mechanisms of S-allele incompatibility in many ways resemble mechanisms of host-parasite incompatibility (30) in that once triggered, the two systems probably use some of the same defense reactions. There is, however, no evidence that the S-allele recognition system serves any function other than that for which it apparently evolved; i.e., for discriminating between selfing and nonselfing pollen, primarily within a species.

The pistil-pollen systems which discriminate between alien species in plants do not usually involve the S-allele system. Pollen of one species frequently germinates and grows poorly on and within the pistil of an alien species, especially in attempted crosses between taxonomically distant species (31). In Hogenboom's terminology (124,125,126), the two partners lack the "congruity" required for one to develop intimately with the other. The pollen of each species has co-evolved with the pistil of its own species and is adapted to it, but is not well adapted to the pistil of other species. The pistil of an alien species may lack the appropriate morphological or chemical features to promote pollen growth, or alien pollen may trigger defense responses in the pistil because it hasn't evolved a way to avoid triggering them. In crosses between closely related species, pollen tubes may develop in pistil tissue but incompatibility is expressed at fertilization or afterwards in the developing embryo (216). In all cases the crosses between species fail because the two are poorly adapted to each other (lack congruity) and not because the pistil recognizes pollen as nonself.

The genetics of species level congruity/incongruity between pistil and pollen are poorly defined but are thought to involve "gene-for-gene" interactions, or (more likely) "genes-for-gene" and "genes-for-genes" interactions between pistil and pollen (125,126). The lack of pistil-pollen adaptation between species is much like the lack of adaptation of fungal pathogens to nonhost species of higher plants, especially if the nonhost is taxonomically distant from the appropriate host (31,130). The pathogen is said to lack "basic compatibility" with the nonhost, and like pollen on an alien species, has difficulty germinating and producing infection structures for entering the plant under attack and may fail to avoid defense reactions of the plant.

TRIGGERING OF DEFENSE RESPONSES

If the defenses of plants to pathogens are not triggered by a general ability of plants to recognize nonself, then how does a plant recognize an invading pathogen and turn on its defenses? Some defenses of the plant are passive devices that are in place before attack and do not require triggering. This is the case for the surface of a nonhost species to which a pathogen is so poorly adapted that it fails to germinate or develop infection structures (130), or for a nonhost which contains preformed substances which are toxic to the pathogen. Active defenses, on the other hand, such as deposition of wall appositions, phytoalexin production, and hypersensitive cell death, are triggered by pathogens which are well enough adapted to start the infection process and are not stopped by preformed barriers or toxins. The available evidence suggests that these defense responses are triggered by certain molecules, often termed elicitors, which are either produced by the pathogen or are released from host structures by the pathogen. (See chapter by Keen). Some elicitors are thought to be located on or within the wall of fungal pathogens where they make early contact with host cells or protoplasts as the pathogens invade. In responding to an elicitor, the plant is not simply responding to a foreign organism, but instead to certain types of molecules produced by the parasite or released by the host in response to the parasite.

The triggering of defense reactions may prove to be more complex than the action of a single elicitor molecule in a given disease, particularly with race-cultivar specificity where the defense reactions are controlled by single corresponding genes in host and parasite and where the controlling mechanisms have not been elucidated. The gene products in gene-for-gene systems may interact directly to elicit defense responses, or may modify the action of general elicitors (Keen, this book; Ellingboe, this book). In either case, the defense reactions in gene-for-gene systems are controlled by interaction of specific molecules of host and parasite.

CONCLUSIONS

I conclude that plants do not possess a general, indiscriminant ability to recognize either nonself individuals or nonself species and that such abilities are not a component of plant disease resistance. The evidence from protoplast fusions and the commingling of tissues in cultures or grafts indicate clearly that higher plants do not generally recognize and reject alien species when cells or tissues are brought into intimate pairings. Some individual species do not tolerate certain others in pairings, but this is the result of characteristics of those particular species and not general recognition of non-self. Likewise, there is no indication that individuals within species are intolerant of other individuals in intimate pairings, except in special cases such as somatic incompatibility between certain strains of fungi.

Higher plants have evolved discrimination systems for selecting sexual mates, including the pistil-pollen S-allele system which forces outbreeding within species of higher plants, or the pistil-pollen discrimination systems which help prevent crossing outside the species. These systems serve only the purposes for which they evolved and do not confer general ability to recognize non-self entities.

The defense responses of higher plants to pathogens are triggered by components and activities of the pathogens and not by recognition of the pathogen as nonself. Implicit in this view is the concept that plants (and other organisms) develop defenses only in response to threats encountered in the course of their evolution. Higher plants have had no need to develop mechanisms at the cellular or tissue level for rejecting either nonself individuals or nonself species. The cell walls and structural surfaces of plants protect them from all but a few highly specialized invaders. Instead of evolving a means of recognizing and rejecting all aliens, the plant has evolved specific mechanisms for dealing with the specialized groups of pathogens it encounters.

IMPLICATIONS OF NONHOST RESISTANCE FOR UNDERSTANDING HOST-PARASITE INTERACTIONS

Michèle C. Heath

Professor, Botany Department, University of Toronto
Toronto, Ontario, M5S 1A1, Canada

To me, an understanding of nonhost resistance is the key to understanding interactions involving plant parasites and their hosts. However, the importance of investigating nonhost resistance is still not widely appreciated. This paper will attempt to highlight the arguments for studying this type of resistance, particularly with respect to plant parasitic fungi. For the purposes of this paper, the terms parasite and pathogen will be used in a broad sense to indicate organisms which can successfully attack at least one species of plant.

WHAT IS MEANT BY NONHOST RESISTANCE?

Nonhost resistance is defined in this paper as the resistance to successful infection shown by a plant species towards a species of plant parasite for which it is not considered a host. To be a nonhost, therefore, all individuals of a species must be resistant to the parasite in question. In practice, this may be difficult to prove, and the recognition of a nonhost normally depends on the "considered" opinion of the researcher. Such an opinion is usually based on the collective observations of generations of plant pathologists and is fairly extensive for the major crop plants of the world. Most plant pathologists would have no qualms in identifying the potato as a nonhost for the wheat stem rust fungus. However, identifying nonhosts among non-cultivated species becomes more difficult because of the enormous numbers of species involved, their genetic diversity, and the relatively little accumulated data concerning their susceptibility to parasites. For example, investigations into the host range of several *Puccinia* species have revealed an unexpectedly large number of wild grasses that may be successfully attacked (10). Even more of a problem is the fact that by chance, collected individuals of a wild plant may be resistant to given isolates of the parasite while other genotypes within the same species may be susceptible. Similarly, it is equally possible that the collected plants may be susceptible to untested genotypes of the parasite while showing resistance to those readily available. The identification of nonhosts is further complicated for fungi that may successfully infect some plants without inducing any visible symptoms (156).

Another theoretical problem in working with nonhost resistance is that one plant species may be "more of a nonhost" than another with respect to a given parasite. At one end of the spectrum are plant species that are genetically and, in many respects, physiologically unrelated to the known hosts of the microorganism in question (e.g. the potato in relation to wheat and the wheat stem rust fungus). At the other end are plant species that are taxonomically related to these known hosts, and which may share with them many genes and physiological processes. If, as suggested later, nonhosts typically have a number of potential defense mechanisms active against a given parasite, then this latter type of nonhost may share some of these with the host species. If these shared defense mechanisms are the ones that the parasite can "overcome" in its host plant, then the nonhost species will have fewer barriers (although not necessarily less resistance) to successful infection than nonhosts at the other end of the spectrum. Even fewer effective barriers may exist in species that are hosts to other *formae speciales* or physiological races of the same parasite species. In my opinion, such plants should not be considered as typical nonhosts. For this reason, the term "species of plant parasite" is used in the definition of nonhost resistance given above. However, this raises yet another problem, because the species concept, as applied to higher plants and animals, is not easily applied to microorganisms. Therefore, the definition and delimitation of a species of plant parasite is often difficult and controversial.

In spite of the caution that must be observed when identifying nonhost plants, there can be no doubt that the host range of any particular species of crop pathogen, even if greater than now perceived, represents only a minute fraction of the more than 250,000 extant species of vascular plants. The remaining species are nonhosts and therefore nonhost resistance is the most prevalent form of plant resistance to disease-causing organisms. It is also the most durable (in the sense of long lasting). In spite of the "jumps" between host species that some pathogens seem to have made during their evolution (202), these seem not to occur to any significant extent in a time span of interest to most farmers and plant pathologists. This is in contrast to the often ephemeral resistance of host cultivars which may last relatively few years in the field (192). To me, these two main features of nonhost resistance, prevalence and durability, provide the basis for arguments as to why this type of resistance should be better understood.

A COMPARISON OF HOST AND NONHOST RESISTANCE

As all plant species are nonhosts to most plant parasites, all are capable of expressing nonhost resistance in some form. Intuitively, one would expect that each plant defends itself in many ways since it is unlikely that any single feature could cope with the diversity of strategies and life-styles found among plant parasites. For example, a thick cuticle might be an effective barrier against parasites that penetrate directly, but is unlikely to be a hindrance for parasites that enter through wounds or stomata. To a certain extent, therefore, what constitutes an effective barrier will vary with the parasite. Moreover, there is no reason why a defense mechanism could not be the absence of a feature required by the particular parasite, rather than the presence of something that inhibits parasite growth. For example, there is some evidence that plants lacking available choline and betaine may be resistant to Fusarium graminearum (217) which requires these compounds. Similarly, some plants that are nonhosts towards certain rust fungi may lack the appropriate topographic surface to encourage urediospore germination and the location of stomata by the germ tube (108,109). However, such reports are outnumbered by those indicating that nonhost resistance is the result of the presence of some inhibitory feature or activity (114,170). Many examples exist where nonhost resistance towards one or more parasites seems to depend, at least in part, on the presence of preformed inhibitory compounds (114,170). Such compounds may have a broad spectrum of activity among parasites, and one suspects that this type of resistance may be more prevalent in non-cultivated species than in cultivated crops since in the latter, such compounds tend to have been bred out (16). As most physiological studies of resistance have been carried out using crop plants, this may explain why there are so many examples where nonhost resistance cannot be completely explained by constitutive features. Often in such cases, treatment of the plants with toxins, heat, or metabolic inhibitors allows subsequent colonization by the parasite (e.g. 111,233); such results indicate that active defense mechanisms are involved.

As discussed by Bushnell (this volume), there is no good evidence to suggest that vascular plants have an indiscriminate "non-self" recognition system. Therefore, there seems to be no reason to believe that the basis of "active" nonhost resistance lies in such recognition, or that the successful pathogen is one which can defeat such a system. Instead, it seems that if a plant's constitutive defenses are not effective against a given parasite, the plant recognizes and responds to a variety of compounds (e.g. fungal wall components) or activities (e.g. plant products released by wall degrading enzymes) likely to be associated with attempted invasion by the majority of potential parasites (64). The alternative theory that a nonhost plant species specifically recognizes some unique feature of each potential pathogen regardless of whether it has previously contacted that organism during its evolutionary history, seems unlikely and I know of no convincing evidence to support it.

Although there are a few reports that link the resistance of host cultivars to constitutive defense mechanisms, the forms of host resistance most often studied seem to involve active defenses (112). In many cases, the typical responses associated with "active" host resistance, such as cell death, phytoalexin accumulation, increased activity of oxidative enzymes, and changes in phenolic metabolism and protein synthesis, are similar to those seen in examples of "active" nonhost resistance (112,114). All may occur in response to a single parasite species and can be elicited by plant-parasitic viruses, bacteria, nematodes and fungi. These responses also bear a close resemblance to those elicited by physical and chemical damage. In many situations, particularly those involving viral infections, it is difficult to envisage, or find experimental evidence, that some of these responses are directly involved in inhibiting further development of the parasite. One explanation of this plurality of responses is that the evolutionary strategy of vascular plants has been towards a certain economy in active defense mechanisms (as in passive ones) so that a variety of agents elicit the same range of responses. Only a few components of this multitude of responses may be effective against any one agent, and any one response may be active against several agents (as suggested by the inhibitory activity of phytoalexins towards insects as well as fungi and bacteria (23)).

In resistant host plants, the control and significance of this multitude of responses have been controversial issues. Although many of the responses may be the result of a "cascade" effect with one process triggering another, it is also possible that some key responses are under the control of a single regulatory gene which is activated by a variety of external or internal stimuli. Such a hypothesis has been put forward to explain how this range of plant responses seems to be controlled by the expression of what classical genetic studies reveal as a single gene for resistance (70). However, the same genetical data also has been interpreted as indicating that all of these responses are secondary phenomena and that host resistance is governed by the direct inhibitory effect of the combined products of the gene for avirulence in the parasite and the gene for resistance in the host (84). Such differing views exemplify how difficult it is at the moment to reconcile physiological studies with classical genetics. Such reconciliation will have to remain in the realm of speculation until molecular geneticists reveal more clearly how eukaryotic genes are regulated, and recombinant DNA technology reveals the nature of the products of the genes for avirulence or resistance. Nevertheless, it is an interesting question whether the similar responses expressed in the same plant when acting as a nonhost or a resistant host are under the same genetic control.

One final point which should be emphasized here is that not only may a nonhost plant have more than one type of defense mechanism against potential parasites, but a single parasite may have to combat several of them in a single plant. For example, with rust fungi, the effective inoculum may be successively reduced by 1) the poor germination of urediospores, 2) the poor recognition of surface features leading to fewer germ tubes locating stomata, 3) the poor growth of infection hypha, 4) the low frequency of haustorium formation, and 5) the death of the haustorium and the invaded cell if a haustorium forms (108,109). Features 1 and 2 seem to be caused by passive defense mechanisms related to the structure of the plant surface, while the rest may involve active responses by the plant. Even if the active responses are controlled by a single gene, it seems unlikely that this gene would also control the surface topography of the leaf. Similarly, the fungal genes that control recognition of the topographical features of the leaf probably differ from those that control the activities or components that trigger active defense mechanisms if the fungus enters the leaf. Thus the genetic basis of this, and probably many other examples of nonhost resistance, is likely to be complex.

Such a conclusion contrasts nonhost resistance with cultivar resistance which often is based, as previously mentioned, on the presence of a single gene for resistance matched by a single gene for avirulence in the parasite (70,82). Whether there is a similar gene-for-gene relationship in nonhost interactions is unknown. Even if nonhost resistance is governed by many genes, it is conceivable that one or more components of nonhost resistance could involve a process controlled by a single gene which is triggered by the product of a "matching" gene in the parasite. Although classical genetic analysis of nonhost resistance is impractical in most instances because of the difficulty in producing interspecific hybrids, the use of mutations in plant and parasite, or the developing techniques of gene manipulation, could help elucidate the genetic basis of nonhost resistance in the future. However, if nonhost resistance involves several defense mechanisms working in concert, changes in a single gene probably will result in a rather subtle change in the interaction between parasite and nonhost. Consequently, such experiments will have to be carried out at the level of biochemical or morphological phenomena, not gross resistance versus susceptibility.

IMPLICATIONS OF NONHOST RESISTANCE FOR UNDERSTANDING HOST SUSCEPTIBILITY

For a pathogen to successfully attack a plant, it must be adapted to successfully recognize and penetrate the appropriate tissue, and to be relatively unaffected by any potential constitutive or induced defense mechanism. What this adaptation entails obviously depends on the potential defense barriers that the growth habit and life style of a particular parasite make it likely to encounter in its host plant. Thus, without an understanding of the nature of these barriers we cannot hope to fully understand the susceptibility of a plant species to its parasites. Experimental evidence suggests that adaptations required by the successful parasite may involve such phenomena as tolerance to preformed plant inhibitors (170), the metabolism of a post-infectionally accumulated phytoalexin (223), the specific suppression of phytoalexin accumulation (185) or the production of a toxin which kills or narcotizes the plant before active defense mechanisms can be expressed (203). Proof of the role of these and other adaptations should be relatively easy to obtain using conditional mutants of the parasite or other forms of gene manipulation (82). However, there are surprisingly few studies in this area although in two of the examples mentioned above, the role of the fungal toxin (203) or enzyme (223) in pathogenesis has been supported by genetic analysis. Similarly, the importance of toxin production in determining host range is nicely illustrated by a genetic study involving Helminthosporium carbonum and H. victoriae (204). These fungi produce host-selective toxins which affect corn and oats respectively. Normal isolates of each species do not infect each other's host; however, crosses between the two fungi resulted in progeny that produced both types of toxin and were pathogenic on both oats and corn.

What is significant in all the examples described above is that the susceptibility of a plant species appears to involve some activity or adaptation on the part of the parasite. This is in direct contrast to the picture of susceptibility provided by geneticists working with susceptible and resistant cultivars of the host plant. Much stress has been placed on the observation that susceptibility of a previously resistant cultivar may be attained by the loss or mutation of the gene for avirulence in the parasite, and the assumed concomitant loss of some active process (82). Similarly, physiological comparisons commonly stress the more rapid and pronounced reactions to infection of resistant cultivars compared with susceptible ones (112). This picture of the parasite not doing something in the susceptible plant when compared with the events in a resistant cultivar supports the idea (82,116) that many examples of cultivar resistance are superimposed on an already achieved "basic compatibility" (82) between the parasite and its host. In such a situation, activities of the parasite that govern successful parasitism of the host species would remain essentially "invisible" in genetic and physiological studies comparing resistant and susceptible cultivars, since such activities would occur in both. This further emphasizes the fact that the key to understanding susceptibility lies in an understanding of nonhost, rather than cultivar, resistance.

If susceptibility of a plant species depends on the parasite having, or being able to develop, appropriate adaptations, then the magnitude and nature of the required adaptive changes will depend on the type of relationship developed between the plant and parasite, and the previous evolutionary history of the latter. In theory, there seem to be three basic routes by which a plant could acquire a "new" successful parasite. In the first (not applicable to viruses), the parasite could have been a free-living saprophyte before it became a parasite of the plant species in question. In such a situation, one would imagine that the primary adaptation required by the potential parasite is the acquisition of those basic features which distinguish saprophytes from parasites. While for some fungi these features may represent the ability to respond to the cues provided by the plant in order to form infection structures, for other fungi and bacteria it is not clear whether the distinction between saprophytes and parasites is anything more than differing abilities of each to overcome plant defense mechanisms. Certainly there is evidence that some saprophytic bacteria can multiply in plants if active defense reactions are inhibited (99) although it seems uncertain whether these are the same defense reactions as those expressed when the plant is acting as a nonhost towards parasitic bacteria.

Regardless of whether saprophytes have to acquire some additional features in order to become potential parasites, all must have adaptations that render "nonhost-type" defense reactions ineffective. Presumably the magnitude of these adaptations would depend on the type of parasitism involved. It might be relatively small for a weak pathogen that infects senescing tissue with poor active defenses, but rather large for a biotrophic organism with a more sophisticated relationship with its host. Whether any biotrophic parasite has developed directly from a saprophyte will be discussed later, but such ancestry might be common for some facultative parasites that are closely related to strict saprophytes. A comparison of the interactions between the appropriate plant species and such related saprophytes and parasites would be of interest. However, few if any such studies have been performed, and the frequency with which parasites are derived directly from saprophytes, or the evolutionary time scale involved, is unknown.

A second route by which a plant species could acquire a new parasite is from an unrelated plant species. While jumps of a parasite between unrelated species have rarely been recorded, taxonomic data suggest that, particularly in the case of rust fungi, such moves must have occurred in the past (202). Given the high degree of physiological specialization that such obligate biotrophs exhibit, it is extremely difficult to explain how a rust fungus, well adapted to one plant species, could suddenly establish a similarly complex relationship with a species totally unrelated to its original host. In an investigation of the interactions between two _Erysiphe_ species (also obligate biotrophs) and a variety of nonhosts, Johnson _et al._ (130) concluded that in plant-fungus combinations other than those involving very close relatives of the host species, the degree to which the fungus developed in a nonhost was more related to characteristics of the plant family than to the taxonomic position of the plant in relation to the host. Similar studies with rust fungi also did not reveal any clear pattern between the degree of fungal development and the taxonomic distance between the nonhost and the host of the fungus in question

(108,109). Significantly, work in Israel has demonstrated that the unexpectedly broad host range of some *Puccinia* species is related to the species composition of the plant communities with which the fungi have coexisted for long periods of time (10). While this observation may be explained by the co-evolution of plant and fungus (discussed below), it is also possible, particularly in the case of alternate hosts for these fungi, that effective nonhost resistance has been "broken" by the accumulation in the fungus of chance mutations during a long-term association of the parasite with the plant. It is also possible that taxonomically dissimilar species may fortuitously have what are perceived by the parasite as similar defense mechanisms that can be overcome by similar adaptations. In such situations, the parasite might require relatively little adaptation to attack a "new" host species if the two organisms were brought into association. This could explain the situation described by Hijwegen (119) where the temperate *Cronartium asclepiadeum* was found to successfully infect tropical plants from several families which it presumably had not previously encountered. Only further comparisons of nonhost resistance in different plant species will determine whether this hypothesis is correct.

The third route by which a plant species might acquire a successful parasite is via a closely related plant species which is already host to the parasite in question. Obviously it is difficult to draw the line between species that are "related" and "unrelated" to any given plant, but one might expect species in the same genus, for example, to have several "nonhost-type" defense mechanisms in common. If a parasite has overcome these in one species, it may be able to overcome them in another. Consequently, resistance in the latter may be based on only a few additional barriers and relatively little adaptation might be needed for the parasite to negate them. Whether it does so presumably depends on whether the plant and parasite are brought into contact, and whether random mutation provides the necessary adaptations. Again, to see whether this principle is valid, more information is needed on the similarities and differences between nonhost resistance exhibited by related and unrelated plants.

A variation of this third route for a plant to acquire a parasite is for the latter to evolve with the "new" host species rather than jumping to it from an already taxonomically distinct relative. Such a route might be expected among obligately biotrophic fungi such as the rust and powdery mildew fungi, and is often supported by taxonomical studies of the fungi (10). Such co-evolution seems to be occurring in the case of the cowpea rust fungus, Uromyces phaseoli var. vignae, and the bean rust fungus, U. phaseoli var. typica. The former attacks cultivated cowpea (Vigna unguiculata) while the latter attacks cultivated beans (Phaseolus vulgaris). It seems likely that both fungi have co-evolved with their hosts from a common ancestral fungus able to successfully attack a plant ancestral to both the cultivated bean and cowpea. Thus, as the plants diverged into different species, so did their parasites. The current inability of each fungus to successfully attack each other's host means either that each plant species has gained some defense mechanism against the inappropriate fungus, or that each fungus has lost the ability to combat one or more defense features characteristic of the inappropriate plant. Interestingly, the resistance shown in each inappropriate plant-fungus combination is expressed before the first haustorium is formed (108; M.C. Heath, unpublished data). In this way, the expression of resistance resembles that seen in nonhosts rather than resistant cultivars of the host plant. Studies are underway to examine the reactions of non-cultivated members of the Phaseolus - Vigna complex to these fungi (86), in the hope that systems such as these may reveal how effective nonhost defense mechanisms, and plant susceptibility, may be gained and lost as both plant and fungus evolve.

One unique way in which an organism may become a parasite of a plant, regardless of its previous history, is to produce a compound toxic to that plant species. Such a toxin effectively negates all active defense mechanisms in one step, rather than requiring the parasite to deal with each one in turn. Such a route to parasitism is exemplified by H. victoriae, which was not originally considered a pathogen of oat plants although it may have been a weak parasite of related grasses (203). However, the introduction into oats of the Pc-2 gene which conferred resistance to some races of crown rust also conferred sensitivity to a toxin produced by some strains of H. victoriae. The way in which the fungus immediately became a serious pathogen of oats suggests that before the introduction of the Pc-2 allele, only active defense mechanisms protected this plant against the fungus. Possibly the fungus was already pre-adapted to any potential passive barriers through its association with wild grasses.

It has been suggested (82) that the H. victoriae - oat interaction exemplifies an early stage in the evolution of parasitism in which a "harmony" between the two organisms is not well established. This implies that the type of interactions controlling susceptibility to a given organism, as well as its genetic basis, may change with time. Such a suggestion echoes previous ones that the more metabolically compatible relationships shown between biotrophic fungi and their hosts develop from less compatible ones involving necrotrophy. Unfortunately, there seems to be little evidence to support such suggestions and, indeed, it has also been postulated that the evolutionary trend may be from biotrophy, through necrotrophy, to saprotrophy (59). If a parasite such as H. victoriae were to evolve into a biotroph, it seems to me that it would either have to forgo its toxin and "start again" with respect to negating nonhost defense mechanisms, or modify the effect of its toxin so that it "anaesthetizes" the cell rather than eventually killing it. Studies which compare the ways in which biotrophs and nonbiotrophs overcome the "nonhost defenses" of their hosts may shed some light on just how different these forms of parasitism are, and how likely it is that one could evolve from the other.

IMPLICATIONS OF NONHOST RESISTANCE FOR UNDERSTANDING HOST RESISTANCE

The control of many plant diseases currently relies heavily on the use of cultivars of the host species which show a resistance to the pathogen that other cultivars lack. In theory, there are two ways in which such cultivar resistance could arise. First, it could be superimposed on an already achieved basic compatibility between plant and parasite as mentioned above. Alternatively, the successful parasite, while achieving a basic compatibility with some members of a plant species, may not be able to achieve this with them all. Conceivably, the latter situation might exist in non-cultivated plants because of their greater intraspecific genetic diversity. This difference in the way in which host resistance is obtained is important since, if superimposed on basic compatibility, the "re-introduced" resistance need not involve the same mechanisms as nonhost resistance; in contrast, at least some nonhost defense mechanisms should still be active if resistant individuals are those which never succumbed to the parasite in question. This latter type of resistance will be termed "residual" nonhost resistance.

Currently, there is not enough information on wild plants to tell which event is more prevalent in nature. However, it appears that plant breeders have "engineered" both types of cultivar resistance in crop plants. An example of the second type, where cultivar resistance relies on "residual" nonhost defenses, seems to be the interaction between Helminthosporium victoriae and oats described earlier. Presumably plants without the Pc-2 allele are resistant to H. victoriae because of the same features which made oats an apparent nonhost of this pathogen before the Pc-2 allele was introduced. However, such resistance (and that shown towards the normally nonpathogenic H. carbonum (233)) is precluded in plants possessing this allele by the death or narcotization of the tissue in the presence of the toxin. Thus toxin sensitivity "overrides" the presence of nonhost defense mechanisms and is inherited as a dominant trait, resulting in the quadratic check (Fig. 1) commonly interpreted as indicating that specificity resides in the "compatible" interaction (82). The lack of development of new races of the fungus that can attack toxin-insensitive plants is not surprising since either a new toxin would have to be produced that would bind to the toxin-insensitive product (if there is one) of the Pc-2 allele or some other way would have to be found to negate nonhost defenses. As suggested elsewhere (116,192), it is to be expected that random mutation would produce new toxins or activities with such specificity at a very low frequency. Thus, this type of cultivar resistance, based on active nonhost resistance that is allowed to be expressed in its entirety because of an insensitivity to the fungal toxin, should be durable. This, indeed, appears to be the case. Interestingly, durable, "partial" resistance to rust fungi in some barley cultivars has been associated with defense mechanisms expressed before the formation of the first haustorium (see 192). Since such defenses are typical of nonhost resistance to rust fungi, it seems possible that in these host cultivars, effective nonhost defenses have also been retained or re-established.

The type of quadratic check seen in Fig. 1 is considered anomalous among examples of host resistance (82) and most plant - parasite interactions give the quadratic check shown in Fig. 2 which is usually interpreted to mean that specificity, involving active features of plant and parasite, lies in the incompatible interaction. Such an interpretation may not be correct (32). However, if it is, resistance could still be based on "residual" nonhost resistance and give this type of quadratic check (Fig. 2) as long as resistance involves active plant responses, controlled by a single gene, that are triggered by a product of the parasite. Such resistance, even though originally a component of nonhost resistance, may not be durable if a random change in the "trigger" (product of the gene for avirulence) can render it unrecognizable by the plant. Random mutations might be expected to produce such non-specific changes at relatively high frequency; as long as these are not

	PLANT SS	PLANT ss
V (toxin +)	COMPATIBLE INTERACTION Plant response to toxin: active nonhost defense reactions inhibited.	INCOMPATIBLE INTERACTION No plant response to toxin: active nonhost reactions effective.
v (toxin -)	INCOMPATIBLE INTERACTION No toxin present: active nonhost defense reactions effective.	INCOMPATIBLE INTERACTION No toxin present: active nonhost defense reactions effective.

Fig. 1. The type of quadratic check exhibited by a host-parasite system where cultivar resistance is determined by active nonhost defense reactions expressed in the absence of a host-selective toxin secreted by the parasite. Shown are interactions of two races of a pathogen differing in their ability to produce toxin and two cultivars of the host possessing the alleles S or s conditioning sensitivity or resistance, respectively, to the toxin.

	PLANT RR	PLANT rr
A	INCOMPATIBLE INTERACTION Recognition of parasite product by plant: active defense reaction(s) triggered.	COMPATIBLE INTERACTION No recognition of parasite product by plant: active defense reaction(s) not triggered.
a	COMPATIBLE INTERACTION No recognition of parasite product by plant: active defense reaction(s) not triggered.	COMPATIBLE INTERACTION No recongition of parasite product by plant: active defense reaction(s) not triggered.

Fig. 2. The type of quadratic check exhibited by a host-parasite system where cultivar resistance is determined either by "residual" nonhost resistance or "re-introduced" resistance that is controlled by a single allele (R) and triggered by the "recognition" of the product of the allele for avirulence (A) in the parasite.

accompanied by a significant loss in parasite fitness, resistance would be short lived.

This type of single-feature, "residual" nonhost resistance cannot be genetically distinguished from that re-introduced after the parasite has achieved a basic compatibility with the whole species, particularly if the same types of defense mechanisms are involved. However, the fact that most types of cultivar resistance in crop plants have been introduced by the plant breeder as single genes into a basically susceptible background suggests that in most cases the "re-introduction" of resistance is the norm. The source of these introduced genes may be other cultivars of the same crop plant, or from different, although usually closely related, species. In the latter situation, it is commonly assumed that it is a component of the donor species' nonhost resistance that is being introduced into the crop plant (although it is rarely determined whether the transferred gene really is involved in resistance in the donor species). However, such introduced "nonhost-type" resistance may be no more durable than that based on "residual", single nonhost-type defenses if it is similarly based on a recognition between plant and parasite that can be easily disturbed by a random change in a component of the parasite (116,192).

From the arguments presented above and elsewhere (114,116), it can be deduced that resistance should be durable if it is 1) multi-component and therefore based on several defense mechanisms, each negated by a different adaptation of the parasite, and/or 2) negated by the production in the parasite of a molecule with a high degree of specificity in its mode of action. Nonhost resistance, therefore, is durable on both counts; in a single plant, it seems to be based on a multitude of potential defense reactions. Furthermore, the way in which a parasite achieves susceptibility with its host species often appears to involve the production of specific molecules that interfere with specific processes in the plant. As described above, certain components of nonhost resistance may not, by themselves, prove to be particularly durable if they can be negated by a nonspecific change in the parasite that is easily attained through random mutation. This may be the case for some putative defense mechanisms that seem common to both host and nonhost resistance expressed by the same species. However, there are instances where nonhost resistance apparently involves different mechanisms from those employed by the same plant in examples of cultivar resistance (108). Conceivably, these mechanisms may be particularly difficult for parasites to negate.

Obviously, more comparisons of host and nonhost resistance are needed to see if these principles are correct. If they are, it should be possible to exploit some of them in producing new resistant cultivars of crop plants, either by conventional breeding or by the more precise techniques of gene manipulation that are currently being developed.

CONCLUSIONS

In my opinion, nonhost resistance is one of the more important, and one of the more neglected, areas of plant pathology. Only by studying nonhost resistance can the features that determine host range and the evolution of parasitism be fully understood. Furthermore, if the principles underlying the durability of nonhost resistance are elucidated, recombinant DNA technology eventually may provide the means by which such principles may be exploited to rapidly produce more long-lasting resistance in cultivars of crop plants.

THE IMPLICATIONS OF GENERAL RESISTANCE FOR PHYSIOLOGICAL INVESTIGATIONS

B. C. Clifford, T. L. W. Carver, and H. W. Roderick

Welsh Plant Breeding Station,
Aberystwyth, Wales, SY23 3EB, UK.

For the purpose of this paper, general resistance is defined as that resistance which is expressed by particular cultivars of a host species to all strains of a pathogen which has evolved a basic compatibility (116,117) with that host species. Such resistance is stable in terms of pathogen variation and thus equates with non-specific resistance, and is durable in agricultural terms. Its alternative - race specific resistance - has well documented attributes that are founded in the gene-for-gene relationship (91). This resistance is under the control of major genes which confer a high degree of incompatibility expressed through the life time of the plant, although there are departures from these qualitative attributes both in terms of genetic control and symptomatology which are more usually associated with non-specific resistance. Genetic control may be complicated by partial dominance, recessiveness, complementary factors and minor modifying genes affecting the degree of compatibility. Expression of resistance may be modified by temperature and by host ontogeny. In the latter case, resistance may increase either with age of tissue (57,179) or age of plant (25,180,212).

It can thus be misleading to infer durability or non-specificity either from host genotype or disease phenotype. What then are the characteristics of non-specificity and how does it differ from specific resistance, if in fact it does? The proof of non-specificty is based on *post hoc* reasoning. At present we are only able to state that a resistance has remained effective as measured by lack of corresponding variation in the pathogen population. In the biotrophic relationships between cereal hosts and both rusts (*Puccinia spp.*) and powdery mildews (*Erysiphe graminis*), no physiological or biochemical attributes have been definitely identified as determinants of non-specificity, and this is the challenge that lies ahead.

Much of the research effort to the present day has centered around race specificity and the biochemical basis of the gene-for-gene relationship: this has so far been unrewarding. Progress in this field may help elucidate mechanisms of non-specificity if only by elimination, but at present we cannot even state that there are different basic mechanisms or, to broaden the discussion, whether they differ from mechanisms of basic compatibility and non-host resistance. Discussions around these issues are found in other contributions to this symposium. For the purpose of this paper, we will assume that in the cereal rusts and mildews, mechanisms governing specific resistance do differ from those governing non-specificity, although the evidence for this is limited and circumstantial. It is largely based on studies of inheritance, symptomatology, and histology, and this has been the approach in our laboratories where oat powdery mildew (*E. graminis* f. sp. *avenae*) and barley leaf rust (*P. hordei*) have been studied in some detail. This paper then is a comparative consideration of the attributes of non-specificity in these systems and is largely descriptive. Attempts will be made to interpret observational data in terms of possible physiological and biochemical control, and it is thus hoped to fulfil the main objective of stimulating further investigations in this field.

In oats, the spring cv. Maldwyn has general resistance to *E. graminis* f. sp. *avenae*. The resistance was described by Jones and Hayes (132) as being quantitative in expression to an increasing degree on each successive leaf. It is thus termed adult plant resistance. Genetic control is through approximately eight additive factors (133). It is environmentally stable, no pathogen isolates with increased virulence have been detected to date, and it has remained effective in the field since 1948 (I. T. Jones, pers. comm.).

The barley system to be considered is exemplified by the cv. Vada which has a non-specific resistance to *Puccinia hordei* derived from *Hordeum laevigatum* (51). It is a non-hypersensitive resistance macroscopically, the expression of which is through the relatively slow development (long latent period) of fewer (low infection frequency), smaller pustules (52). It is expressed in the seedling stage but to a greater degree on flag leaves (188) and results in 'slow rusting' in the field. The genetic control is through a single partially dominant gene together with four or five minor genes (189) and, although adapted pathogen isolates have been detected in glasshouse tests (55) and a field trial (190) it has remained effective in widespread commercial use in northwest Europe since 1969. It has been designated 'Type II resistance' by Clifford (54) and will be referred to as such in this paper.

During the course of the infection cycle, both *P. hordei* and *E. graminis* go through an extra-cellular or pre-biotrophic phase and an intra-cellular or biotrophic phase (Fig. 1). In the rust, the former phase consists of spore release, flight and landing, germination, ingress through the host stoma, sub-stomatal vesicle formation and the production of primary infection hyphae and haustorial mother cells. The equivalent processes in the mildew are simpler as penetration of the host epidermal cell is effected directly via the appressorial germ tube. In both organisms, the biotrophic phase begins with penetration into the host cell and production of an haustorium. Subsequent development is by external hyphae and intracellular haustoria in *E. graminis* and by inter-cellular hyphae and intra-cellular haustoria in *P. hordei*. Conidiospore production and release completes the cycle. Host resistance, which may be

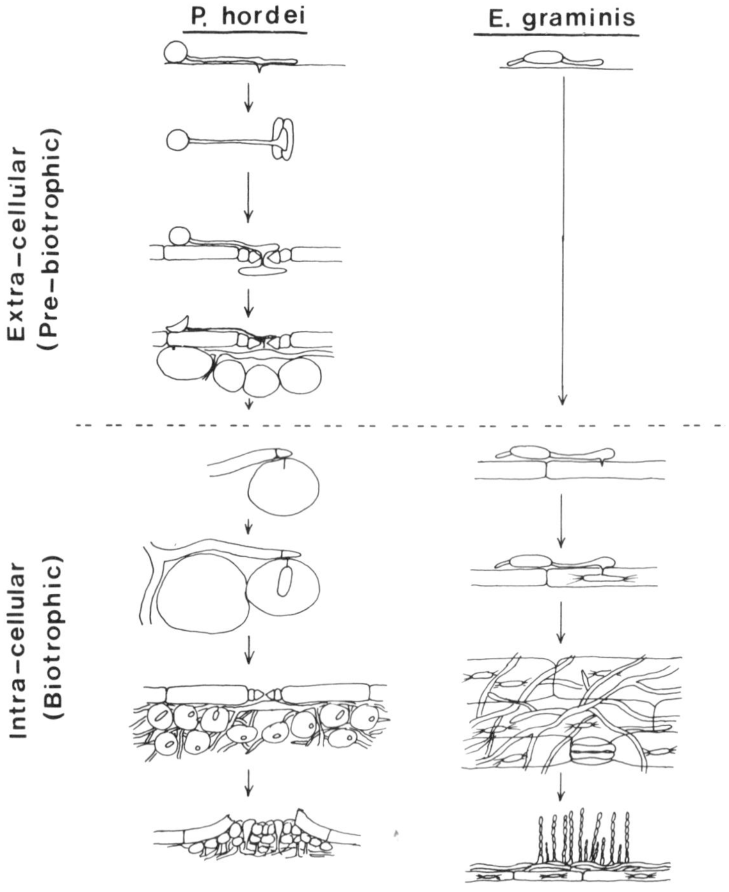

Fig. 1. Schematic representation of the asexual infection cycle in Puccinia hordei and Erysiphe graminis.

defined as any host-mediated event which results in a restriction in the infection cycle, may conceivably operate at any phase in the cycle. The following is a consideration of these host-mediated barriers as they may relate to general resistance mechanisms.

PRE-BIOTROPHIC PHASE

Germination and Germling Development

Erysiphe graminis appears to be unique among the powdery mildews in that it alone germinates to produce a germling with two germ tubes whereas other species form only one (154). The first, or primary germ tube (145) remains short, is aseptate, and forms neither an appressorium nor an haustorium. It emerges from the conidium soon after release from the mother colony and generally becomes attached to the host surface soon after its emergence. Primary tube emergence appears to require no stimulation other than release of the condium from the mother colony. Carver (39) found that short tubes, like the primary tube, were produced by conidia suspended on microthreads of small orb-weaving spiders. Multiple short tubes were formed frequently, but appressorial germ tubes were not observed. Such conidia transferred to coleoptiles dissected to leave a single layer of epidermal cells (34), quickly responded by producing appressorial germ tubes. The stimulus to form appressorial tubes was not provided by stasis, or by contact of the conidium with glass or agar on which only short or apseptate longer tubes were formed. Hence, a particular property of the host is perceived by the germling, possibly through the primary tube. When conidia germinate on host tissue, the primary germ tube often attaches to the host surface before emergence of the appressorial tube, and in doing so it engenders host responses such as the appearance of an autofluorogen (143,155), the congregation of a transient cytoplasmic aggregate (153), and the eventual deposition of a papilla beneath the tube tip (148). As early as 2 h after inoculation an infection peg can be formed in the contact surface beneath the primary tube tip, and a corresponding pore is visible in the host surface (152). From this it is clear that physiologic contact can be established between host and primary tube, although it is not absolutely essential for the primary tube to become firmly attached for recognition to occur (39,40).

It is not possible to rule out the involvement of chemicals in stimulating normal appressorial development, but Yang and Ellingboe (230) concluded that only the host's wax layer was important and that its physical properties were more important than chemical constituents. From this it seems less likely that a germling could detect physical properties of waxes through its relatively thick conidial wall than through the tip of the newly emerged primary germ tube. The detection of these properties may be achieved by simple contact by the primary germ tube and not require its firm attachment to the host surface.

As well as its probable role in very early recognition of the host, the primary germ tube can play a vital part in supplying water from the host to the germling during appressorium differentiation under arid environmental conditions during germination (40). To fulfill this role the primary tube must be firmly attached to the host surface. Under humid incubation conditions, appressorial growth can occur even if the primary tube is not firmly attached; here the germling's water requirements may be provided by the release of metabolic water or by uptake of atmospheric water through the conidial wall or the walls of the germ tubes (40). The levels of certain elements such as Ca and Si become elevated in conidia soon after primary tubes attach (151), and the vital dye acridine orange can be taken up at about this time (149). It seems most likely that these components are taken up through the primary tube, but there is no evidence to show whether host materials other than water are of direct benefit to the germling, and it seems possible that they are taken up incidentally as solutes along with water.

The available evidence clearly points to the importance in germling development of the primary germ tube as a structure involved both in early recognition of the host and in obtaining water from it under dry atmospheric incubation conditions. Any host character that interferes with either process is likely to confer resistance of a race non-specific nature. All race specific gene-for-gene interactions studied so far require, as a prerequisite to their expression, physical or chemical interaction between host and pathogen; thus gene-for-gene incompatibility has never been seen to be expressed in cereal mildew until infection pegs are produced from appressoria (81,129). Resistance which limits primary germ tube association with the host would have to act to prevent interaction. This might be achieved through preformed host characteristics such as leaf hairiness (127), abnormal wax structure (230), cuticle thickness (100), or wall silicification (97).

It is interesting to consider briefly here the behavior of powdery mildew conidia on non-host species which do not normally support this pathogen. Johnson, Bushnell and Zeyen (130) compared barley mildew development on non-host Graminae, Liliaceae and Iridaceae. On graminaceous species not reported to host any powdery mildew, rates of appressorium formation were slightly lower than on barley. On species of Liliaceae and Iridaceae, less than 10% of conidia formed appressoria as compared to 70% on barley. Failure of germlings to recognize the non-host surface may well have contributed to their failure to form appressoria.

In the germination of cereal rust urediospores, no primary germ tube is formed and germination occurs only in the presence of free moisture. A number of germ pores are present and these would appear to function in the absorption of water and as an egress for the germ tube, only one of which is normally produced per spore. The germling gains ingress to the host following production of an appressorium and penetration peg over the stomatal opening. A number of physical and physiological barriers may operate to restrict this phase of development. In wheat it has been shown that urediospores of Puccinia recondita are trapped more effectively by cultivars with high numbers of leaf hairs (Fig. 2) in wind tunnel experiments. Conversely, it may be argued that the presence of such hairs limits intimate contact between fungus spore and host plant surface which could, for the reasons discussed above, have a negative effect on spore germination especially in the case of E.

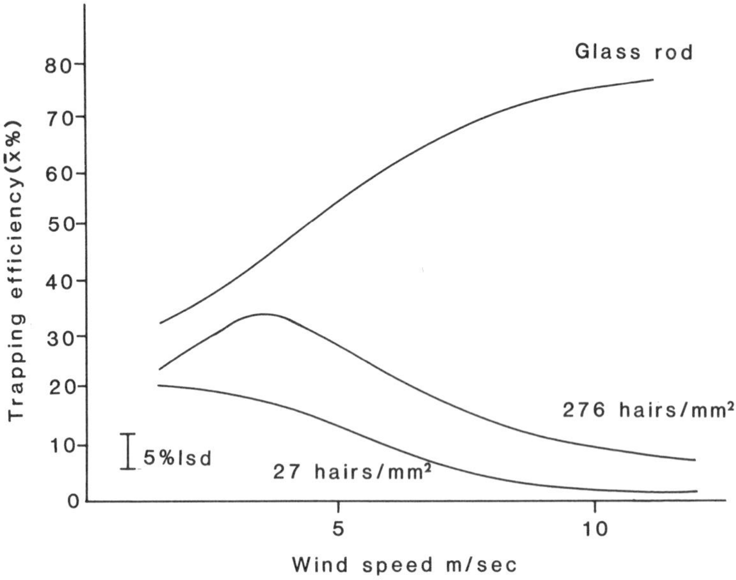

Fig. 2. Trapping of Puccinia recondita urediospores by wheat cultivars differing in leaf hair numbers. (B. C. Clifford, unpublished data)

graminis.

In P. hordei it has not been determined whether germ tube development is in response to thigmotropic or chemotropic stimuli. Germ tube orientation is generally at right angles to the long axis of epidermal cells indicating a physical response to leaf structural characteristics. Schematic representations of the range of germination patterns are depicted in Fig. 3. It is tempting to speculate that stomatal recognition and appressorial formation can only be perceived by mature germ tubes and that this is a chemotropic response of which the young germ tube is not capable. However, the pattern of responses depicted suggests that this may be an oversimplication. Although the majority of tubes grow at right angles to the long axis of the leaf, indicating a thigmotropic response to leaf surface structure, others do not. Some appear to seek out the stoma and produce appressoria from short germ tubes; others grow

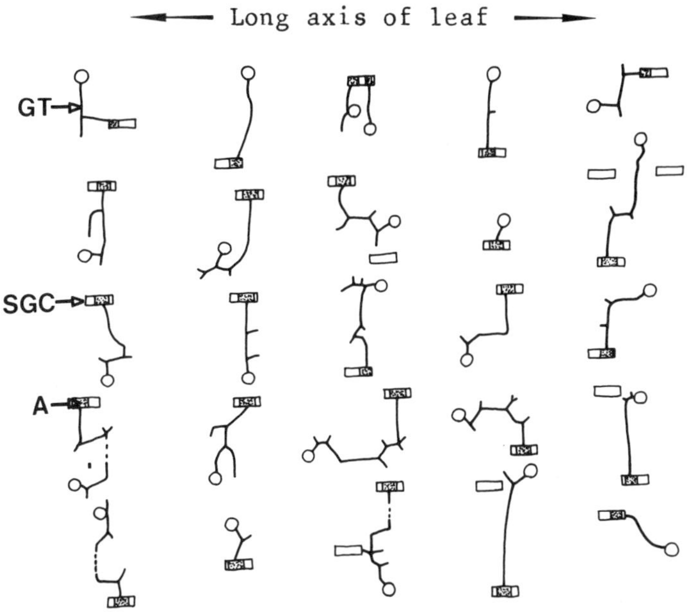

Fig. 3. Schematic representation of uredial germling development of Puccinia hordei on adaxial leaf surfaces of the spring barley cv. Midas. (H. W. Roderick, unpublished data). (GT = germ tube; A = appressorium; SGC = stomatal guard cells) (H. W. Roderick, unpublished data)

past stomata appearing to ignore them. Clearly, there is the possibility of non-specific barriers operating which would be interesting to study. A more obvious barrier is the stoma itself both in terms of numbers and distribution. Observations in our laboratory (B. C. Clifford and H. W. Roderick, unpublished data), have shown that seedling leaves of the barley cv. Peruvian have approximately 30% fewer stomata compared with cv. Gold and this is reflected in a corresponding reduction in numbers of infections (Table 1). Clearly, such a barrier is non-specific.

The next stage of germling development in P. hordei is the production from the penetration peg of the sub-stomatal vesicle (SSV). This is a cigar-shaped structure which is normally oriented with the

Table 1. Numbers of penetration events for Puccinia hordei, race F, on cvs Gold and Peruvian and host stomatal numbers* (B. C. Clifford and H. W. Roderick, unpublished data).

Parameter	cv. Gold	cv. Peruvian
Stomata numbers	3629.3	2533.5
Penetrations (after 2 days) (%)	74.8	55.1
Penetrations (after 4 days) (%)	76.2	55.8
Pustule numbers (after 10 days) (%)	62.6	37.2

*Per cm^2 of seedling leaf

long axis of the stoma (Fig. 4). The vesicle illustrated shows an abnormal orientation but was chosen as it clearly depicts the structure unobscured by stomatal guard cells. In an histological study,

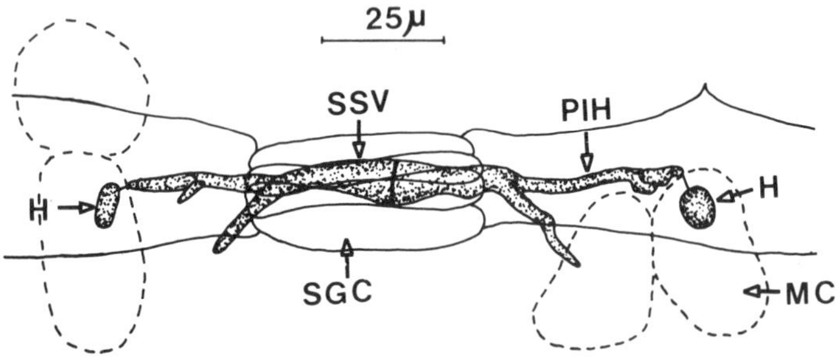

Fig. 4. Schematic representation of sub-stomatal vesicle and primary infection structures of the uredial stage of Puccinia hordei. (H. W. Roderick, unpublished data). (SSV = sub-stomatal vesicle; PIH = Primary infection hypha; H = haustorium; MC = mesophyll cell; SGC = stomatal guard cell).

which compared cv. Vada with the highly susceptible cv. Midas (52) no quantitative or qualitative differences between the cultivars were found in germling development prior to SSV formation. Germination, germ tube development, appressorial formation and stomatal penetration proceded normally in cv. Vada as did the production of SSVs (Table 2). Resistance began to be expressed visibly with development of the fungal thallus from the SSV. Normally, primary infection hyphae develop from both ends of the SSV (bipolar development) and these terminally differentiate an haustorial mother cell which effects penetration of the host mesophyll cell via the infection peg. This in turn forms the primary haustorium to initiate the biotrophic phase (Fig. 4). In cv. Vada, significantly more SSVs failed to develop normal bipolar primary infection hyphae 48 h after inoculation compared with cv. Midas (Table 3) and significantly more vesicles produced none or only one infection hypha. As there is as yet no intracellular contact between host and pathogen it may be inferred that this expression of resistance relates to extra-cellular recognition processes, physical or chemical, that have been discussed above for earlier

Table 2. Development of primary infection structures of Puccinia hordei on cvs Vada and Midas 24 h after inoculation [data from Clifford (52)].

Event	Cultivar Midas	Cultivar Vada	p^t
Sub-stomatal vesicles	83.4*	77.7	NS
% Appressoria forming SSV	96.3	95.7	NS

*Mean number of events per 100 microscope fields x 10 replicates

p^t = probablility in t-test of paired comparisons

NS = not significant

Table 3. Development of primary infection structures of *Puccinia hordei* on cvs Vada and Midas 48 h after inoculation [data from Clifford (52)].

Event	Cultivar Midas	Cultivar Vada	P^t
Sub-stomatal vesicles	11.6*	22.7	≤ 0.01
Primary infection hyphae			
i) Monopolar	21.0	27.8	≤ 0.01
ii) Bipolar	63.5	47.9	≤ 0.01

*Mean number of events per 100 microscope fields x 10 replicates

P^t = probability in t-test of paired comparisons

phases of the pre-biotrophic stage.

ESTABLISHMENT OF THE BIOTROPHIC STAGE

Primary Host Cell Penetration

The next easily recognizable phase of development at which host resistance may operate is penetration of the host cell, and this has been extensively investigated for *E. graminis*. Penetration is achieved through the production of an infection peg which emerges from a pore in the underside of the appressorial lobe. The peg is blunt-ended and appears to have no cell wall at its tip (176). Although it has been suggested that cuticle penetration is achieved mechanically it seems likely that wall penetration has an enzymatic component (2,29,145). For over a century it has been known that attempted penetration can elicit a host response (213) and this is now recognized as being extremely complex (5,33,237).

Even before penetration of the cuticle and host cell wall by the infection peg, ultrastructural changes in epidermal cell apoplast structure occur, induced perhaps by chemical communication (33,176, 201). Fluorescence microscopy reveals an autofluorescent material, of similar appearance to that seen beneath primary germ tubes, in the host wall beneath the appressorium (143,155). By light microscopy, host wall changes are seen as a 'halo' effect in the host cell around the tip of the appressorium (3,166,201). In living cells, the cytoplasm of the cell under attack is seen to be mobilized and an aggregate forms beneath the appressorium (28,33). This cytoplasmic aggregate is involved in apoplast changes including the formation of a paramural papilla through a system of vesicle-mediated material deposition (5,237). For many years it has been suspected that papillae may be involved in resistance to fungal penetration (1), although the correlation between the presence of papillae and resistance to penetration is not absolute; papillae are often penetrated and, following penetration, papilla constituents may continue to be deposited in the haustorial neck collar (4,147,237). Many workers have attempted to correlate the numerous and variable physical and chemical characteristics of papillae with resistance to penetration, but it has always been impossible to prove a definite cause-and-effect relationship.

Carver and Carr (41) compared oat mildew development at the primary infection stage on 10 lines of oats with various levels of resistance to mildew. These included cv. Manod (highly susceptible hexaploid control), cv. Maldwyn (quantitative adult plant resistance), three highly resistant genotypes of wild oat species, and various hybrids between these species and cv. Manod. They demonstrated a strong negative correlation ($\underline{r} = -0.843$, $\underline{P} < 0.001$) between the numbers of germlings arrested with visible infection pegs embedded in a host papilla, and the number which formed a colony. Even in the suscept it was at the papilla stage that many penetration attempts failed. In the seedling leaf no more germlings were arrested at the penetration stage on cv. Maldwyn than on the suscept, but the adult plant resistance of cv. Maldwyn was manifested on leaf 5 by a significantly reduced proportion of germlings that succeeded in breaching papillae. The wild species also expressed resistance which limited penetration beyond the papilla and, although this resistance was strong even in seedling leaves, their hybrids with the suscept showed adult plant enhancement of resistance similar to that in cv. Maldwyn. It is not clear whether the reasons for failure to penetrate papillae were the same in all genotypes, but it is clear that polygenic, stable resistance can prevent intrusion of a papilla. Another expression of resistance limiting colony formation was found in the wild oat species where a small proportion of germlings appeared to penetrate the papilla structure, the tip of the infection peg swelled to form an incipient haustorium, but there was no further development and the fungal structure became encased by thickened membranes. Hirata (122) noted a similar phenomenon in barley but it is not known whether this encasement is a separate response from papilla deposition, or merely an extension of it.

The phenomenon of resistance to penetration by E. graminis has interested researchers for many years and much of the descriptive work in the literature is centered around the papilla response. In barley, Lin and Edwards (161) showed that where penetration attempts failed, papillae contained a 'basic staining material' that was never present in papillae that were penetrated successfully. Their implication was that there was a qualitative difference between the two types of papillae. However, they were unable to ascertain whether the material evident in some papillae conferred the resistance to them, or whether the material was produced following arrest of the attempt for some other reason. Although it is quite possible for qualitative differences between host response sites to account for success or failure of penetration, it seems equally likely that quantitative variation between sites e.g. in the amount of a particular component incorporated into a papilla structure, could affect the outcome of a penetration attempt. The association of increased rate of silification of papillae with failure of attempted penetration serves as an example of this type of quantitative variation.

Silicon has been implicated in resistance to infection by *E. graminis* since Germar (97) reported reduced mildew infection in wheat plants grown in soils supplemented with silicon dioxide and suggested that this was due to silicification of 'Membranen' preventing the ingress of haustoria. Hirata (122) reported a similar effect when barley leaves were floated on nutrient solution containing soluble silicon, but he attributed the resistance to 'callus' (papilla?) material surrounding the infection structure. Kunoh and his co-workers were the first to demonstrate localized accumulation of Si in epidermal cell haloes and papillae associated with *E. graminis* attack (146,147,148,150). They demonstrated that Si first appeared 12 h after inoculation and accumulated not only in encounter sites where penetration failed, but also in papillae that were penetrated and remained as 'collars' around the haustorial neck. Limitations inherent in their technique prevented Kunoh's group from defining the role of Si in resistance to penetration. To overcome these limitations, Zeyen *et al.* (238) devised a procedure for relating cytoplasmic detail of powdery mildew infection to the presence of insoluble Si by sequential use of light microscopy, scanning electron microscopy (S.E.M.), and X-ray microanalysis (Fig. 5). Using this procedure, Carver *et al.* (47) studied rates of Si accumulation in contact sites of barley leaves attacked by genetically virulent mildew germlings. In confirmation of previous reports, Si first appeared in host tissue beneath the first appressorial lobe at 12 h when papillae were also first visible. By 16 h the level of Si in the site had increased substantially. At 20 h three distinct germling phenotypes were distinguishable: those which had definitely succeeded in penetrating and had formed an haustorium; those which had definitely failed to penetrate from the first appressorial lobe and as a consequence had differentiated a second lobe; and thirdly, indeterminate germlings which could not be described as successes or failures, having formed neither an haustorium nor a second appressorial lobe. Comparision of numbers of Si X-ray counts (Table 4) obtained from papillae

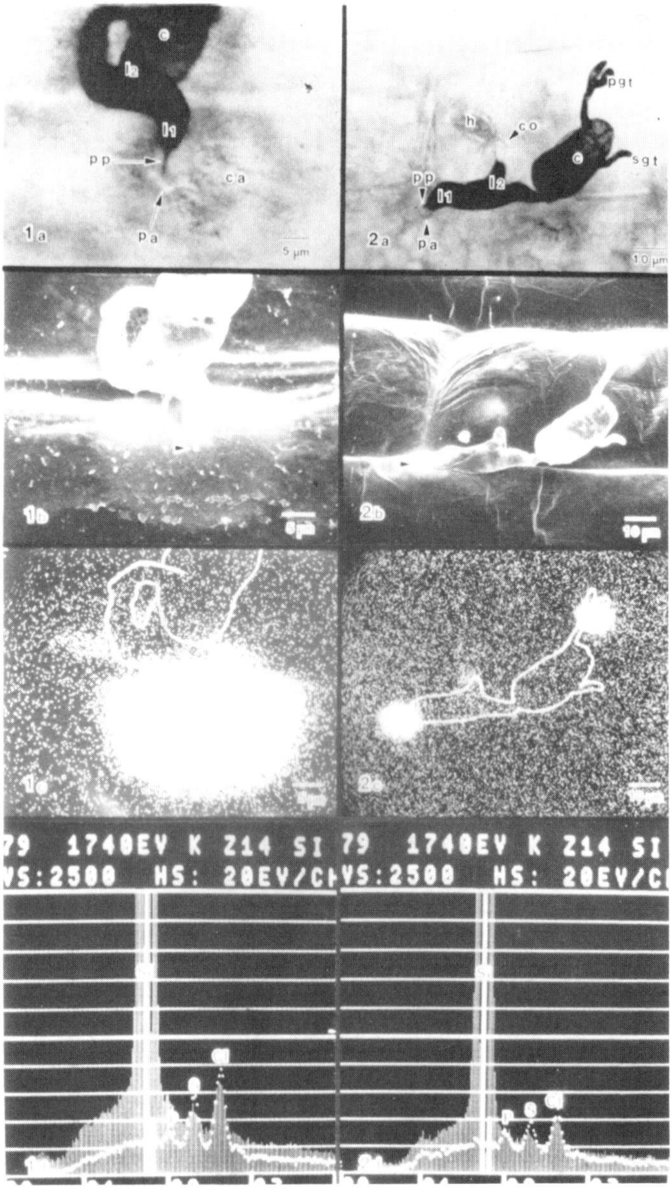

Fig. 5. Relating cytoplasmic detail of powdery mildew infection to presence of insoluble silicon by sequential use of light-microscopy, SEM, and X-ray microanalysis.

Plate 1. _E. graminis_ germling 20 h after inoculation where attempted penetration from the first appressorial lobe failed, leading to differentiation of a second appressorial lobe. (a) Light micrograph (tissues stained with Coomassie Brilliant Blue R-250). (b) SEM, secondary electron image, surface topography. Arrow indicates where a stationary electron beam was positioned to obtain the bar spectrum in (d). (c) SEM, X-ray dot map (400,000 dots) showing distribution of Si around the contact site, with outline of germling positioned for reference. Note, Si was distributed in the entire SEM 'halo' area and not restricted to the papilla. (d) Two, super-imposed X-ray spectra collected for 100 sec each. Vertical bar spectrum from stationary beam positioned over papilla as in Plate 1(b); white dot outline spectrum from an area of the host cell at least 100 μm from the papilla-aggregate site. A comparison of the two spectra using net peak areas showed no real differences in P, S, or Cl. There was a 500x increase in net peak area for Si at the papilla aggregate site; the intensity of the Si peak, white centroid marker, exceeded the available vertical scale of the monitor.

Plate 2. _E. graminis_ germling 24 h after inoculation where attempted penetration of the first appressorial lobe failed. A second appressorial lobe differentiated and succeeded in penetrating and establishing a young haustorium. (a) Light micrograph. (b) SEM, secondary electron image. Arrow indicates the position of a stationary electron beam used to take the vertical bar spectrum in (d). (c) X-ray dot map micrograph (40,000 dots) for Si showing an intense signal from the primary germ tube and first appressorial lobe contact areas, but no such intensity from the haustorial collar area. (d) X-ray spectra collected under the same conditions used in Plate 1(d); details as in Plate 1(d). Again, comparison of net peak areas for Si between the appressorial germ tube contact area and unaffected sites reveal a 485x increase in Si, but no real differences in P, S, or Cl.

Abbreviations: c, conidium; ca, cytoplasmic aggregate; co, haustorial neck collar; h, haustorium; l1, first appressorial lobe; l2, second appressorial lobe; pa, papilla; pp, penetration peg; pgt, primary germ tube; sgt, subsidiary germ tube. [From Zeyen, Carver, and Ahlstrand (238)].

Table 4. Mean intensities of Silicon X-ray peaks associated with host tissue close to the first appressorial lobe of successful and unsuccessful attempted primary infections by E. graminis hordei attacking genetically compatible barley [data from Carver, Zeyen, and Ahlstrand (47)].

Time after inoculation (h)	Unsuccessful attempts (second appressorial lobe formed) counts	Successful attempts (haustorium formed) counts
20	95,583	33,405
24	86,629	82,641

associated with the definite failure and definite success phenotypes showed far higher levels of Si associated with the failures. At 24 h the same three germling phenotypes could be distinguished, but by this time Si levels associated with penetrated papillae, left as haustorial neck collars, had increased dramatically and were similar to Si levels in papillae beneath failed penetration attempts. The level of Si associated with failures did not increase after 20 h indicating that the Si deposition response had been discontinued at this site. The simplest interpretation of these data is that there is a dynamic interaction between host and pathogen in which the intruding infection peg must penetrate the papilla before sufficient Si is deposited in it to arrest the attempted intrusion, probably by mechnical strength and resistance to enzymatic degradation. As with all other studies that have correlated papilla characteristics to the success or failure of penetration, the work on Si deposition fails to demonstrate a definite cause-and-effect relationship, but the observations, together with the known physical and chemical characteristics of insoluble Si, support the view that silicification may impart resistance. However, even if Si does play such a role it certainly is not the only factor influencing penetration: in one of the leaves examined by Carver *et al.* (47), there was no detectable accumulation of Si in contact sites even though some penetration attempts had definitely failed and large papillae were associated with them.

It is tempting to speculate that host genotype as well as cultural conditions (97,122) might influence positively the availability of Si to, or its potential rate of deposition at, an infection site. If this is the case, the resistance conferred to such genotypes is likely to be of a general, race non-specific nature. This hypothesis might be examined using the technique of Zeyen *et al.* (238) to look at host genotypes expressing this type of resistance.

If one considers the morphological and physiological structure equivalent to the mildew appressorium in the rusts, namely the haustorial mother cell (HMC), a number of interesting parallels may be drawn. In a light microscope study of the cv. Vada:*P. hordei* system, Niks (181) reported that approximately 25% of establishing colonies failed to develop beyond the HMC stage in what he termed 'early abortion' implying that resistance operates on contact between HMC and host mesophyll cell. Chong and Harder (50) made a study of *P. coronata avenae* HMC development in oats using transmission electron microscopy which showed the HMC to be structurally specialized. Specialization was first seen as the deposition of additional wall layers in the hyphal tip associated with the host cell wall and they suggested that host:pathogen recognition may occur at this point. In a further paper, Niks (182) reported that in cv. Vada, unsuccessful penetration attempts by uredial colonies of *P. hordei* were manifested as lobed infection hyphae suggesting repeated penetration attempts following failure. Powdery mildew appressoria, too, can develop multiple lobes if attempted penetration by the first is unsuccessful (21,47,134). Niks also observed host cell depositions at the point of penetration similar to the papilla-response to mildew. Similar papilla-like thickening of host cell walls associated with HMCs together with the formation of a penetration peg was described by Harder and Chong (105) but for normally-compatible interactions. They concluded from ultra-structural evidence that penetration is a wall dissolution process (162) and that wall deposition could act as a mechanical barrier. Niks (181) concluded that 'early abortion' of P. hordei colonies resembles non-host resistance and that the genes in cv. Vada governing the response are part of the system of basic incompatibility. Part of the argument was based on the premise that major gene-governed hypersensitivity operates after haustorial formation. This is certainly not the case in powdery mildew systems where many major genes prevent haustorium formation (81) although the expression of resistance does require physiologic contact (81,129). A further parallel with the mildew system comes from the studies of Heath who observed silicon deposition at the point of contact between HMCs of the cowpea rust fungus (Uromyces phaseoli var. vignae) and cells of French beans (110). She later demonstrated silicon deposition in necrotic cowpea cells infected by an incompatible isolate of the cowpea rust fungus (115). In addition, she showed (113) that suppression of silicon deposition in French bean leaves resulted in increased numbers of successful penetrations and haustorium formation by the bean rust fungus. This suggests a causal relationship between Si deposition and host resistance.

The central question is - what governs success or failure? Are we observing the same manifestation of different underlying mechnisms or of a single underlying mechanism? Certainly, known race specific resistance, governed by major genes, can be quantitatively expressed. We have temperature sensitivity, ontogenetic variability and, most intriguingly, the X-reaction. The latter case is, at least superficially, very similar to cv. Vada's Type II resistance to P. hordei. In it, a range of reaction types may occur on a particular host plant carrying a specific gene for reaction to a genetically uniform population of pathogen propagules. The proportion of events of a particular reaction type may vary with temperature or with homozygosity of the corresponding loci in the rust and host plant. This may confound the issue which is that the X-reaction may be expressed where the pathogen is homozygous for avirulence and the host is homozygous for resistance. Mechanistically, all these situations can be explained by interacting gene products on a gene-for-gene basis with the outcome depending on the quality and quantity of product formed.

A similar situation is found in the powdery mildews and, curiously, is exemplified by the reaction of cv. Vada (which possesses the mildew resistance Mlv) to mildew isolates which lack corresponding virulence. In this case sporulating colonies do develop, but they are less vigorous than those produced by a fully virulent isolate, and the underlying host tissue shows necrotic discoloration. Microscope studies (46) showed that host cells could respond in one of three different ways to attempted infection: i) papillae formed, infection failed, but the host cell remained alive; ii) the host cell showed very rapid hypersensitivity and died before haustorium formation; iii) the host cell accepted the infection, a primary haustorium developed completely normally, and colony growth proceeded as in a fully compatible relationship. In the last case, subsequent attempts to penetrate the host cell followed establishment of the young colony. If these attempts challenged the cell containing the primary haustorium, the attempt succeeded and a secondary haustorium was produced. If, however, the attempt challenged a neighbouring cell, the cell either accepted the penetration, or it responded with hypersensitive collapse (Fig. 6). The distribution of these 'susceptible' and 'resistant' cells appeared to be random across the leaf, they did not occur in stripes, as would be expected with a chimaera, nor was there any apparent influence of proximity to stomata or vascular bundles. The difference in the reactivity of neighbouring cells may be due to genetic differences, or alternatively to phenotypic differences which affect the cells' ability

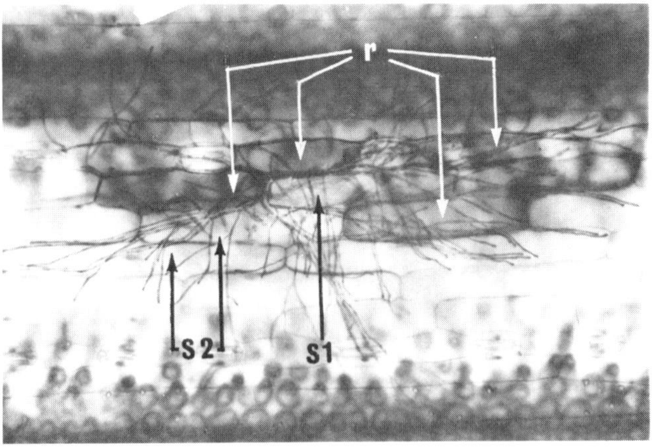

Fig. 6. Colony of E. graminis f. sp. hordei, isolate AB3, attacking cv. Vada which shows a mosaic of 'susceptible' and 'resistant' epidermal cells. S1 = susceptible cell containing primary and secondary haustoria; S2 = susceptible cell containing secondary haustoria; r = resistant, collapsed cell, stained darkly with aniline blue. [From Carver and Williams (46)].

to respond to the fungal intrusion.

Many studies of powdery mildew have revealed variation in the susceptibility of different epidermal cells to attempted penetration, with cells near stomates frequently being more susceptible than the larger ones further removed from the stomates (e.g. 100,123,129). The variation is not due to differences in the cells' capacity to react hypersensitively, but it might relate to the strength of the papilla response. Stomatal subsidiary cells are often found to be exceedingly prone to infection, and in some highly resistant genotypes almost all colonies that establish depend on primary penetration into subsidiary cells.

In mildew, some hosts are so resistant to penetration that as few as 1% of germlings are able to establish colonies by 'normal' means. In these cases it is often found that a majority of the colonies that do establish, form an initial relationship by penetrating stomatal subsidiary cells. Hirata (122) first noted preferential subsidiary cell infection in barley but, although this is also seen in some genotypes of oats, the frequency of penetration of subsidiary cells in other resistant genotypes does not exceed the statistically expected frequency (41). This suggests that subsidiary cells are not necessarily more susceptible than others, but their obvious susceptibility in some instances still awaits explanation. One other means of by-passing epidermal cell resistance to penetration deserves mention. In a study of various barley genotypes derived from old European cultivars and land-races, Carver (38) found that colonies could be established by endophytic infection. This was first reported by Salmon (200) but has rarely been mentioned since then. In Carver's study, endophytic infection accounted for most of the colonies found on some highly resistant lines. Here, the occasional conidia that were deposited above, or very close to stomata, sometimes germinated so that their appressorial germ tube grew down into the sub-stomatal cavity, and elongated until it encountered a mesophyll cell. An appressorium differentiated against the cell, and a haustorium formed within it. Hyphae were then able to ramify within the leaf where other haustoria were formed in other mesophyll cells. Hyphae also emerged from the mother conidium to grow across the leaf surface, and these too were then able to form further haustoria in the epidermis. This showed that even though the gene(s) for resistance were highly effective in limiting primary infection of the epidermis they were not expressed in the mesophyll. Furthermore, once endophytic infection was established, the epidermal resistance no longer functioned. This might be because the established colony is more energetic than the germling, or because mesophyll infection had induced susceptibility of the overlying epidermis.

Differences in response between cell types have also been reported for avirulent isolates of P. graminis f. sp. tritici on wheat carrying gene Sr_6. Invaded mesophyll cells autofluoresced but epidermal cells did not (199). As Rohringer and Heitefuss (198) point out, this difference is of interest because in up to 40% of the infection sites the first haustorium is formed in an epidermal cell (211). One assumes, perhaps naively, that cells of the different host tissues are genetically the same but they clearly differ physiologically.

Haustorial Development and Colonization

The next definable phase of development is the production of haustoria and the growth of the mycelial thallus. The haustorium is critical to the establishment of the biotrophic relationship between pathogen and host, and it is generally accepted that the prime function of the haustorium is nutritional.

The haustorium of E. graminis is a complex structure with a number of bipolar digitate processes (Fig. 7) which increase its absorptive area. At first sight the D- (or dikaryotic) haustorium of Puccinia spp. appears to be a relatively simple structure lacking any form of processes which could increase its absorptive potential. Powdery mildew is of course totally dependent on its haustoria for nutrient uptake whereas the rust fungi may absorb nutrients directly through their intercellular hyphal cell walls making them less dependent on haustoria and obviating the need for a complex structure. However Harder and Chong (105) have recently re-examined the features of D-haustoria and these are illustrated in Fig. 8. Of particular interest is the tubular complex which appears to connect the extrahaustorial membrane to the host endoplasmic reticulum (Fig. 9). It may well be that this connection facilitates the access and efficient passage of nutrients from the host cell into the haustorium and movement of fungal materials into the host cell.

Returning to the cereal mildew system, host resistance governed both by single genes and polygene systems can affect the development of cereal mildew after penetration of the host cuticle/cell wall barrier. In discussing studies of infection by E. graminis, Ellingboe (83) drew two conclusions

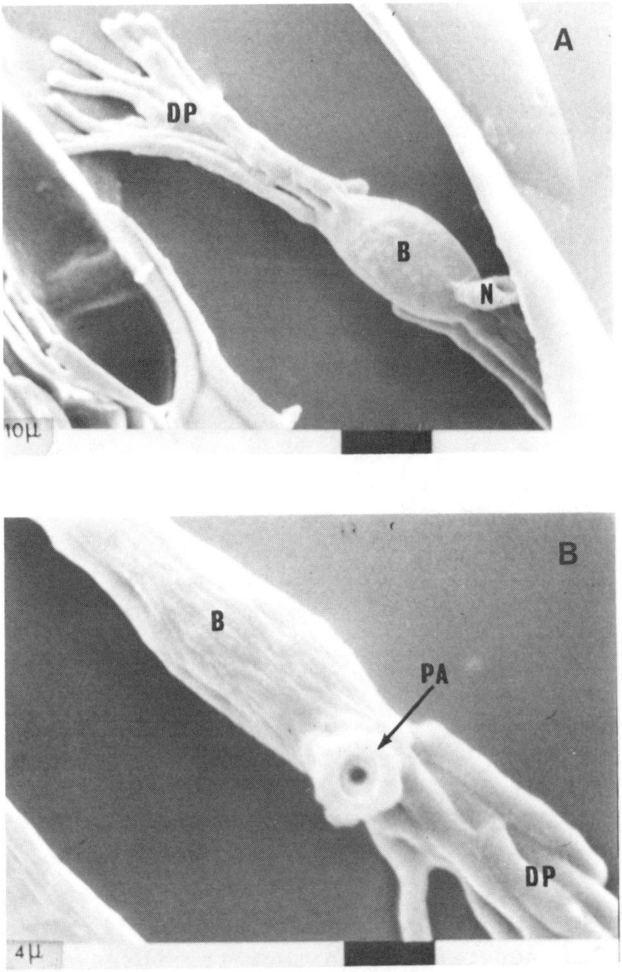

Fig. 7. Scanning electron micrographs of oat mildew haustoria. B = haustorial body; DP = digitate processes; N = haustorial neck; PA = host papilla remaining as collar around haustorial neck. [From Carver and Chamberlain (44)].

regarding the phenotypic expression of incompatibility conditioned by a gene-for-gene interaction: 'each incompatible parasite/host genotype affects the ontogeny of interaction in a unique way, and each has

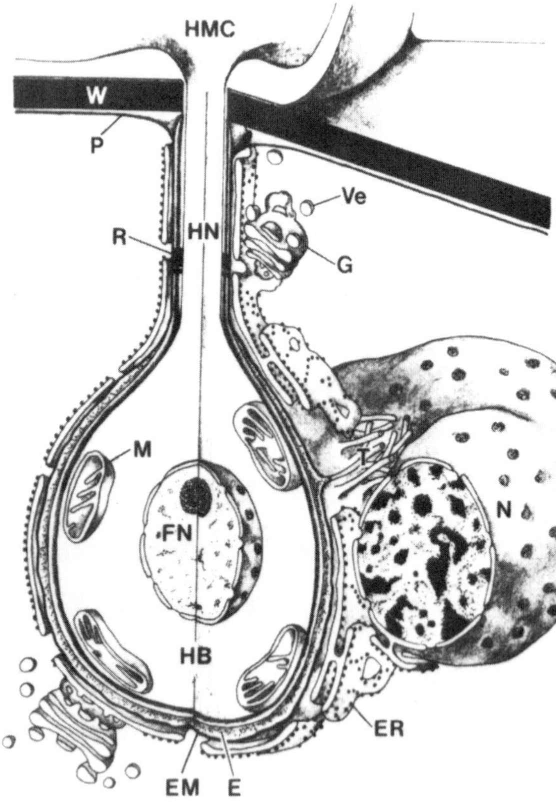

Fig. 8. Diagram of an invaded host cell cut open at the site of penetration to show the three-dimensional structure of a mature D-haustorium of P. coronata, and its association with the host cell organelles involved. The structures are not drawn to scale, and some are illustrated by only a few examples (e.g., Golgi bodies, vesicles, ribosomes). E, Extrahaustorial matrix; EM, extrahaustorial membrane; ER, endoplasmic reticulum; FN, fungal nucleus; G, Golgi body, HB, haustorial body; HMC, haustorial mother cell; HN, haustorial neck; M, mitochondrion; N, host nucleus; P, plasmalemma; R, neck ring; T, tubule complex; Ve, vesicle; W, host cell wall. [From Harder and Chong (105), reproduced with permission.]

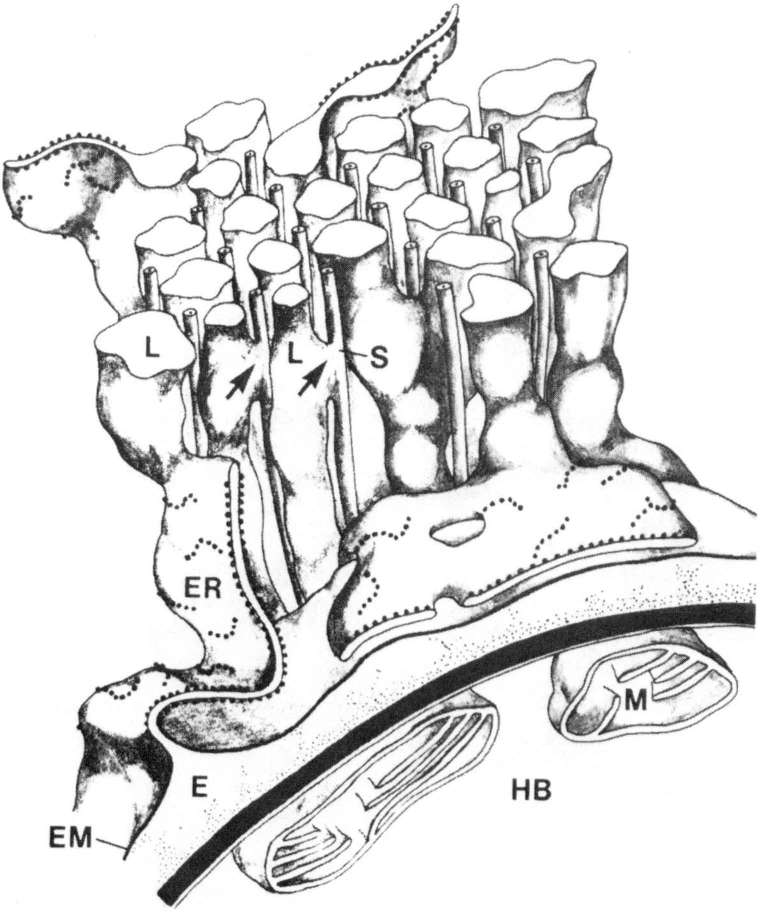

Fig. 9. Diagram of a haustorium (P. graminis f. sp. tritici)-associated organized membrane complex (reconstructed from a series of serial sections) cut open to show the principal components and their interrelationships. Note the connections (arrows) between the large (L) and small (S) tubules. The large tubules are also connected to the surrounding host endoplasmic reticulum (ER), which in turn is continuous with the extrahaustorial matrix (E). EM, extrahaustorial membrane; HB, haustorial body; M, mitochondria. [From Harder and Chong (105), reproduced with permission.]

several different effects, on a time scale, during the ontogeny of interactions between a host and parasite'. Thus in interactions where incompatibility is due to a single host resistance gene, the majority of parasite units may be arrested at a particular developmental stage, but others may be arrested earlier or later, and some may develop fully to produce sporulating colonies. The proportion of parasite units arrested at each stage depends upon the particular genotypes involved in the interaction. Similarly, in genetically compatible interactions, a large number of parasite units form vigorous sporulating colonies although some fail and die at earlier stages. Hence in terms of the phenotypic expression of interactions, the difference between incompatibility and compatibility might simply appear to be a modal shift within a continuous distribution of interaction types from one extreme to another.

In cases where the infection peg penetrates the host cell wall and papilla successfully, the tip of the infection peg starts to swell into an incipient haustorium usually between 16 and 24 h after inoculation, depending upon host and pathogen genotype and favorability of the enviroment. Our current knowledge on the nature of the further development of the host-parasite interface is discussed in a recent extensive review by Manners and Gay (168). The host plasmalemma is not punctured by the intruding infection peg, but invaginates around the expanding haustorium to form the extrahaustorial membrane. Between this membrane and the haustorial cell wall lies the extrahaustorial matrix. The extrahaustorial membrane appears to be derived from the host plasmalemma but is significantly different from it in many ways. The transition between normal plasmalemma and modified extrahaustorial membrane occurs abruptly at a point of attachment of the membrane to the haustorial neck in the region known as the neckband. A seal is formed in the neckband between the fungal cell wall and plasma membranes and this effectively isolates the extrahaustorial matrix from the general apoplast of the leaf. It can be reasonably concluded that all nutrients necessary for colony growth must pass from the mesophyll through the leaf apoplast to the epidermis, into the cytoplasm of the infected cell, through the extrahaustorial membrane and matrix before finally passing the haustorial cell wall into the fungus. Having established the haustorial structure, nutrients start to flow into the superficial fungal structure and secondary hyphae start to form at about the time that digitate processes emerge at each end of the now elipsoidal haustorial body (28,43,121).

In a compatible interaction, the next few days see the accelerating growth of the young colony which draws nutrient through the single, but enlarging, primary haustorium (43,122). The primary haustorium grows for 4-5 days, increasing its surface area by the production and elongation of digitate processes (43,120). For 2-3 days colony growth is restricted to the production of hyphae, but after a certain threshold amount of growth is exceeded, and this is approximately 25 hyphal cells (195), a second generation of haustoria is produced from the hyphal weft. This second generation is produced synchronously and in response to darkness in the diurnal cycle (43,45,46,122). It is not known whether darkness influences the pathogen directly or indirectly via its influence on host metabolism. During the following light period, growth is again directed to hyphal production until the next dark period when a third haustorial generation is produced. This growth rhythm appears to be maintained as long as the colony continues vigorous growth. Conidiophores are generally first produced shortly after the third haustorial generation, but their production is arrhythmic as is the subsequent formation of conidia.

Carver and Carr (42) observed a number of restrictions on normal haustorial development in the resistant cv. Maldwyn in comparison with a susceptible cultivar. The overall length of mature primary haustoria was reduced as was the number and total length of digitate processes. Further calculations indicated a corresponding decrease of approximately 11% in surface area with the implication of a reduction in absorptive potential.

Limitation of size does not appear to be the only factor affecting nutrient uptake, as indicated by studies of young colonies fixed prior to the formation of secondary haustoria i.e. when all the growth of the colony is represented by secondary hyphae (42). At this stage, the primary haustorium is immature and it is relatively easy to assess its size accurately by measuring the total length of digitate processes produced by an haustorium. Similarly, it is relatively simple to assess the quantity of nutrient taken up by the haustorium by measuring the length of secondary hyphae produced by the colony. Analysis of variance was used to compare these fungal characters on different hosts. When various oat genotypes were compared in this way, it was found that, although resistance of some lines did reduce the size of the immature haustoria, and this was associated with reduced hyphal growth, the haustoria found at this stage in cv. Maldwyn were significantly larger than those in the suscept although no more hyphae were produced. This indicated that the efficiency of haustoria, in terms of their ability to extract nutrient for hyphal growth, was impaired in cv. Maldwyn. To test this further, regression analyses were performed where length of digitate process/haustorium was regarded as the independent variable, and length of secondary mycelium/colony as the dependent variable. The regression coefficients ('b' values) obtained (Fig. 10) may be regarded as measures of haustorial efficiency. From the susceptible control genotype a 'b' value exceeding 3.0 was obtained; from the wild oat species <u>Avena barbata</u> the 'b' value was less than 1.0; from cv. Maldwyn the 'b' value was approximately 2.0. Analysis showed that these values differed from each other. Thus host resistance can affect nutrient uptake by haustoria without affecting their size.

Quantitative resistance acting to restrict nutrient uptake for whatever reason can delay achievement of the threshold colony size necessary for response to darkness and initiation of secondary haustorial production (195). The number of haustoria produced in each generation, subsequent to the

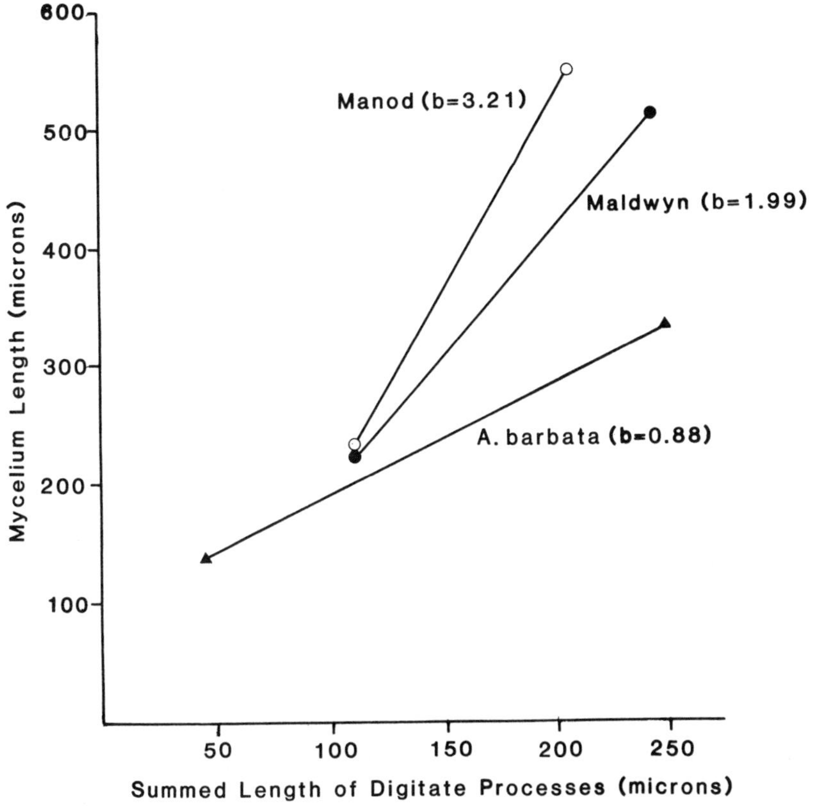

Fig. 10. The relationship between primary haustorium size and mycelium production after 52 h in the fifth leaf of oats. [From Carver and Carr (43), reproduced with permission.]

primary, can be reduced (Table 5), although the effect on numbers of secondary haustoria is not nearly as pronounced as on numbers of tertiary haustoria. Presumably this is simply because small differences in relative susceptibility of hosts, which act continuously during colony development, become more obvious as time passes and they are compounded. Counts of numbers of tertiary generation haustoria provide an extremely sensitive measure of colony size which can be used to detect small differences between levels of

Table 5. The mean number of haustoria produced per colony in leaf 1 (seedling) and leaf 5 of cv. Manod (susceptible), Maldwyn (adult plant resistant), and *Avena barbata* (highly resistant genotype) [data from Carver and Carr (42)].

Number of haustoria/colony		Host genotype			
		cv. Manod	cv. Maldwyn	*A. barbata*	D_{5}%*
Total (including	Leaf 1)	9.77	9.53	4.53	1.72
one primary)	Leaf 5)	8.50	6.30	3.36	
Secondary	Leaf 1)	2.27	3.37	1.43	0.78
	Leaf 5)	3.03	2.20	1.25	
Tertiary	Leaf 1)	5.50	5.17	2.10	0.99
	Leaf 5)	4.47	3.37	1.10	

*D values apply for comparison of leaf means within host genotypes, and of genotype means within leaves.

host resistance relatively easily (42). This measure is particularly useful because it gives an indication of the future potential for nutrient absorption by the colony as well as the compatibility of the past relationship.

Similar observations on the effects of Type II resistance on haustorial development have been made for the *P. hordei*:cv. Vada system. In considering these we would like to re-introduce the question of the mechanistic relationship between specific and general resistance. We will argue that the following evidence supports the case for there being separate mechanisms governing these resistances in the barley leaf rust system. The cultivar Cebada Capa carries the gene Pa_7 (184) which conditions a 'hypersensitive' interaction with avirulent isolates of *P. hordei*. In an histological study of this interaction in comparison with the highly susceptible cv. Gold and cv. Vada, Clifford and Roderick (56) observed that although avirulent isolates of *P. hordei* penetrated and colonized cv. Cebada Capa, this development was arrested at the point of sporulation at which time the relationship was terminated: this being manifested by host cell death and visible necrotic pin-head sized lesions. Quantitative aspects of colony development in cv. Cebada Capa were similar to cv. Vada as measured by colony size and numbers and conidiophore production (Fig. 11). Development was much reduced compared with cv. Gold. In addition, haustorium numbers per colony were similar in cvs. Cebada Capa and Vada (Fig. 12) as was the relationship between size of colony and numbers of haustoria (Fig. 13). The authors concluded that, although the final outcome of the interaction in cv. Cebada Capa was governed by gene Pa_7, colonization was controlled by a different mechanism similar to that of cv. Vada. Support for this comes from the genetic study of Parlevliet (191) who concluded that cv. Cebada Capa carries a number of minor genes conditioning partial resistance in addition to gene Pa_7. The above observations substantiate the earlier report (53) of a close correlation between the numbers of compatible and incompatible infection sites resulting in different

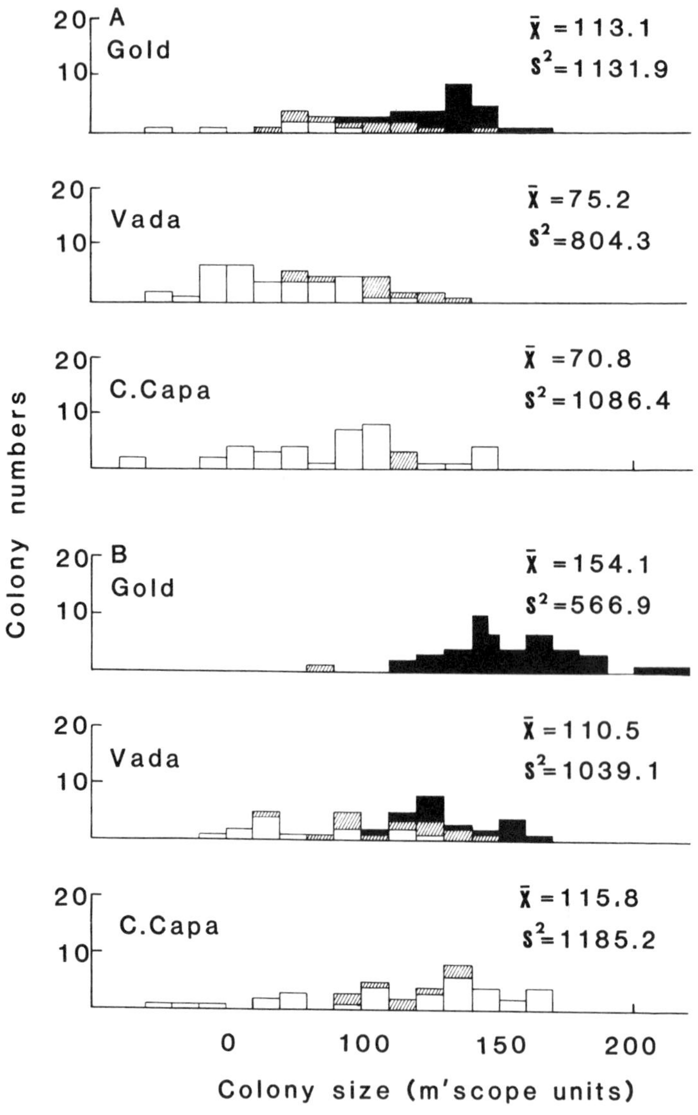

Fig. 11. Growth and development of populations of P. hordei race F colonies on three spring barley cultivars (open bars = vegetative stage; cross hatched bars = conidiophore stage; solid bars = sporulating stage.) (A) 6 days after inoculation, (B) 8 days after inoculation. [From Clifford and Roderick (56), reproduced with permission.]

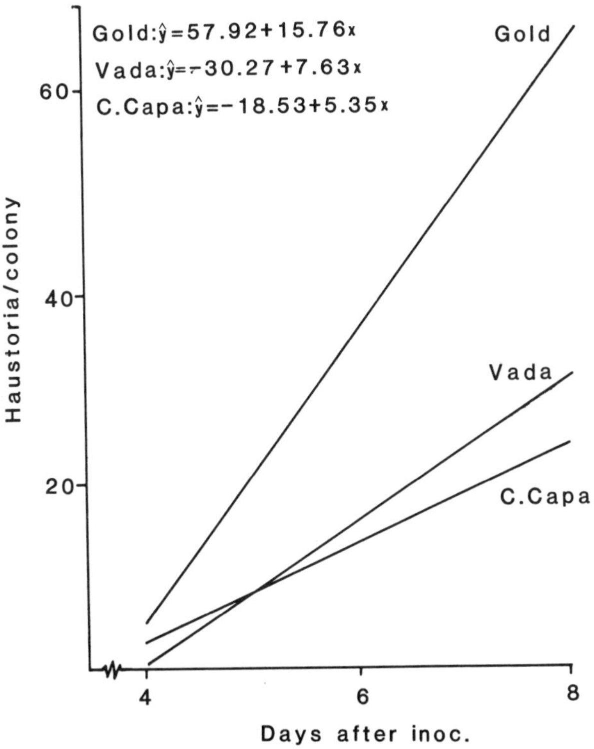

Fig. 12. Haustorial development in colonies of P. hordei race F in three spring barley cultivars up to 8 days following inoculation. [From Clifford and Roderick (56), reproduced with permission.]

Pa_2-gene carriers when challenged with either virulent or avirulent isolates of P. hordei. This partially-expressed 'background' resistance can also be detected where the major gene is temperature-sensitive: Mayama et al. (174) observed the same number of autofluorescing sites at 20°C (incompatible) and at 26°C (compatible) in wheat plants carrying gene Sr_6.

Because of the role of the haustorium in the nutrition of the rust and the observed effects of the cv. Vada-type resistance on haustorial development, it is of interest to examine the relationships further. In the mildew system we saw that not only were the

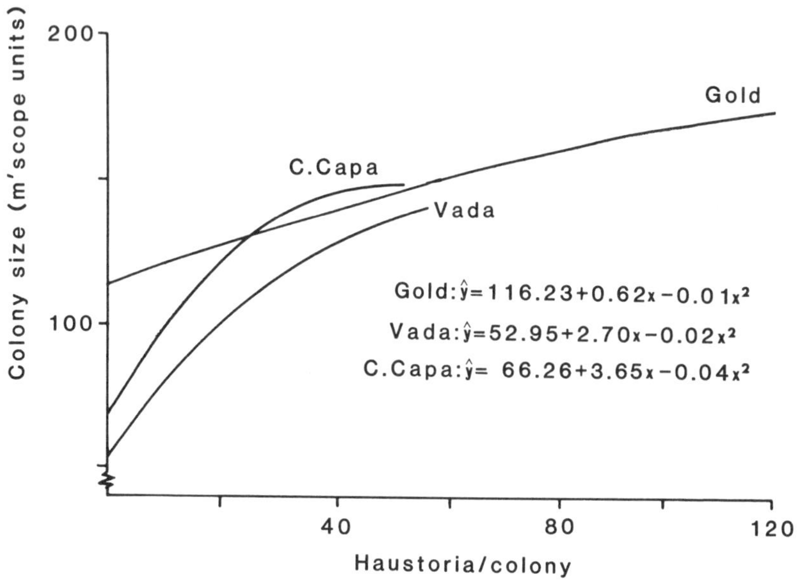

Fig. 13. Relation between colony size and haustoria numbers of P. hordei race F in three spring barley cultivars 8 days after inoculation. [From Clifford and Roderick (56), reproduced with permission.]

size and number of haustoria reduced in cv. Maldwyn but that the efficiency of the haustoria in supporting mycelial growth was also affected. A recent study in our laboratory (B. C. Clifford and H. W. Roderick, unpublished data) examined this relationship in the barley brown rust Type II resistance system. Numbers of haustoria per unit colony area of P. hordei on cv. Ricardo, which has partially-expressed resistance similar to that of cv. Vada, were fewer than in the highly susceptible cv. Gold (Table 6) when assessed on colonies in the vegetative, conidiophore and sporulating stages. At six days after inoculation, colonies were larger on cv. Gold and the majority were producing conidiophores: these relatively more mature colonies had more haustoria per unit amount of mycelium. It is tempting to suggest from these limited data that the haustoria have a specialized nutritional role which relates to sporulation,

Table 6. Numbers of haustoria per unit colony area of *Puccinia hordei*, race F, on barley (B. C. Clifford and H. W. Roderick, unpublished data).

Stage of colony development	Time (d)		
	6	7	11
cv. Gold			
Vegetative	3.58(15)*	2.56(10)	-
Conidiophore	4.20(25)	3.32(11)	-
Sporulating	-	3.34(18)	3.84(39)
Unclassified	(0)	(1)	(1)
cv. Ricardo			
Vegatative	3.13(35)	2.74(19)	3.85(8)
Conidiophore	3.24(5)	2.66(15)	-
Sporulating	-	2.67(6)	4.57(31)
Unclassified	(0)	(0)	(1)

*Average of (n) events from a total of 40 observations.

Table 7. Numbers of colonies per cm^2 of *Puccinia hordei* on barley 10 days after inoculation (average of 5 reps.) (B. C. Clifford and H. W. Roderick, unpublished data).

Colony stage	Gold	Ricardo
Vegetative	1.5 ± 0.9	8.2 ± 2.6
Sporulating	96.2 ± 22.2	53.5 ± 20.8

possibly in providing nutrients and materials specifically relating to sporogenesis. In colonies where haustorial production and/or function is impaired there is a consequent impairment in sporogenesis. It is appreciated that experiments such as the one described above provide only a crude quantitative evaluation. Measuring colony area is inadequate to describe three-dimensional colonies. Inadequacies of staining and observational techniques make haustorial assessments difficult. It is now possible to make assays of fungal biomass using biochemical markers such as mannan or chitin (226) which can give a much greater degree of precision. More exciting is the prospect of *in vitro* and *in vivo* studies of haustorial function to examine the nature of biotrophy and host-mediated restrictions on it.

Sporogenesis

Although early abortion can account for some of the expression of Type II resistance to *P. hordei* it has been shown that resistance continues to operate in reducing and delaying colonization. This is finally expressed as delayed and reduced sporulation (52) and is further illustrated in Figure 11. Some mycelial colonies fail to sporulate in the normal time span of the relationship as observed for cv. Ricardo (Table 7) where, 10 days after inoculation, the total number of colonies was approximately two thirds of those on cv. Gold and of those, 5% were in the vegetative stage as opposed to only 1.5% in cv. Gold (B. C. Clifford and H. W. Roderick, unpublished data). The sporogenetic area is also smaller as seen visibly by relatively small pustules but this does not necessarily relate to correspondingly reduced mycelial tissue. Indeed, Whipps *et al.* (226) showed that, although the sporulating region was much reduced in colonies in cv. Peruvian (Type II resistance) many of these colonies were as large or larger than in cv. Gold.

Why then is this so? We can speculate that the key lies with haustorial function. In both the rust and mildew systems, haustoria are reduced in numbers, size and efficiency although there is no information to suggest how host genotype affects this. It might be through an influence on the plasmalemma which is modified as it invaginates around the intruding infection peg and haustorium. In mildew, it may be that the efficiency of transport of nutrient from mesophyll symplast to leaf apoplast to epidermal symplast varies between diseased genotypes. It may simply be that less utilizable nutrient is produced by some genotypes.

Haustoria may have a specialized role in obtaining specific metabolites necessary for sporulation. This may be the provision of sugar alcohols such as mannitol, arabitol and erythritol or in the sequestering of polyphosphates. Carbohydrates certainly accumulate at infection sites and the ratio of mannan: chitin increases at sporulation (226). It would be helpful to analyze spore compositions in this regard. This is obviously an area inviting future research.

General Consideration and Conclusions

There are many stages in the ontogeny of fungal biotrophy when host resistance can operate to impede development. In the *P. hordei* : cv. Vada and *E. graminis* : cv. Maldwyn systems we have identified some of these barriers, namely reduced penetration of host cells and consequent prevention of haustorium formation, and restriction in haustorium size and efficiency resulting in lower infection frequency, slower colony development and reduced and delayed sporulation. These resistances have been stable and effective over a considerable period of time (cv. Vada since 1969; cv. Maldwyn since 1948), and no pathogen isolates capable of overcoming them have been selected in the agricultural situation; they are apparently race non-specific (54, I. T. Jones pers. comm.) and appear to have their basis in restricting basic compatibility. Both resistances are governed by complex genetic systems (133,189).

Several hypotheses have been advanced to explain the molecular basis of compatibility and resistance and, although the effects of chemical interaction may be expressed while penetration is being attempted, i.e. before haustorium formation is achieved, the hypotheses do require physiologic interaction between host and pathogen. Heath (116,117, this book) summarized the significance of 'basic compatibility' in relation to host/parasite specificity. She considered that a parasite able to infect a particular host plant species must have the capacity to avoid or negate the consequences of general defenses that make most plant species resistant to most plant pathogenic species. Such a parasite is 'basically compatible', with its host at the species level of specificity. Specificity at the host genotype/pathogen genotype level can only be determined after the establishment of basic compatibility. Bushnell and Rowell (32) suggested a model to explain the relationship between 'non-host' resistance, basic compatibility, and the type of resistance conditioned by single host genes to specific fungal races. They suggested that specific receptor sites exist in the host and if these fit specific suppressors produced by the fungus, then the recognition prevents the operation of defense reactions which would otherwise have been triggered by non-specific fungal elicitors: this relationship leads to compatibility and overrides 'non-host' resistance. If the host then acquires a specific resistance gene which alters the receptor site so that the fungal suppressor no longer fits, the elicitor is free to trigger the defense reaction. Compatibility is restored if mutation to virulence alters the specific fungal suppressor so that it once more fits the receptor. It is not clear how polygenic resistance reconciles with this model. It is possible to visualize situations whereby the host receptor site may be modified through the action of several, or many, genes each having a minor effect, so that the goodness of fit between fungal suppressor and receptor is impaired, and its restoration requires mutation of corresponding virulence genes of minor effect. However, it seems equally likely that polygene resistance of a quantitative type could operate through an entirely independent but superimposed system, or systems, and that the resistance genes do not interact directly with pathogen genes concerned with virulence. This form of resistance would not be affected by the level of specificity, so that the level of resistance conferred by the genes would not be affected by the virulence genes possessed by the pathogen. In this sense the resistance would be race non-specific. However, this does not preclude the possibility that modification of the genetic background of the pathogen might erode polygene resistance through selection of genes for increased 'aggressiveness'. In this sense pathogen aggressiveness corresponds to non-specific resistance in the host. It is a character that has nothing to do with establishing basic or specific compatibility through the production of suppressors, but it modifies the pathogen's general ability to grow on a particular host phenotype. In this way the pathogen may become specifically-adapted to overcome, to a greater or lesser extent, the quantitative resistance of a host. In this context we would wish to avoid the inference that individual genes in a polygenic system are necessarily controlling specific components of resistance that would relate to different barriers in the sequential events of infection. As an alternative, each minor gene may additively affect the basic mechanism of resistance which may be expressed through a restriction of pathogen development at any stage of the infection process. As yet it has not been possible to identify and associate specific minor genes with specific barriers to infection but it would be of great interest to do so.

It is reasonable to suppose that basic compatibility is an evolving and thus quantitatively expressed state, which is achieved through various host:pathogen interfaces of both a biophysical and biochemical nature. It is also reasonable to assume that the various host characteristics which limit or allow basic compatibility are under various genetic controls. It is equally unreasonable to suppose that quantitative variation in the expression of basic compatibility is under any particular genetic control e.g. additivity, recessiveness etc. What is crucial is the host plant character, physical or chemical that is in play. It seems axiomatic to the authors that progress in understanding the nature of biotrophy and of host restraints on it will best be achieved through coordinated genetic, cytological, physiological and biochemical investigations. It is regrettable that much of the information, including our own, has been obtained from studies of existing host genotypes which have already assembled genetic systems. This may confuse the issue in that part of the genetic system may be governing one resistance mechanism and another part an entirely different mechanism. One can envisage an assembly of genes controlling recognition systems, toxin production, hormonal balance, nutritional relationships, etc., each confusing the identification of the other. Plants with such resistance complexes are likely to be selected both naturally and artificially and, although durability of resistance is likely to be achieved through such a process, it makes the investigator's task that much more difficult. It also makes it that much more challenging and exciting.

PROGRESS IN UNDERSTANDING THE BIOCHEMISTRY OF RACE-SPECIFIC INTERACTIONS

N. T. Keen

Department of Plant Pathology, University of California, Riverside, CA 92521.

The molecular basis of gene-for-gene specificity in plant-pathogen interactions is one of the most important unsolved questions in plant pathology. As Ellingboe has persuasively noted (84), genetic evidence strongly suggests that the molecules which convey recognitional specificity in these systems are the products of dominant plant disease resistance genes and pathogen avirulence genes. Despite this simple genetic basis and the availability of host and parasite lines differing at the respective resistance and avirulence gene loci, isolation of the gene products has not yet been accomplished. Fortunately, progress has occurred in the last few years and, most encouraging, molecular cloning techniques are now available that promise to greatly accelerate the research. This paper will review the current status of investigations on the basis of gene-for-gene specificity and discuss new experimental approaches. In my terminology, a resistant plant reaction corresponds to an incompatible plant-pathogen interaction, while a susceptible plant response results from a compatible interaction. Race-specific resistance is involved with gene-for-gene interactions, while general resistance, sometimes called non-pathogen resistance, is a broader form, usually effective against an entire pathogen species, pathovar or *forma specialis*.

Plants may employ one or a combination of executional mechanisms for expression of single gene resistance to incompatible pathogen races. Virtually all known disease resistance genes modulate the hypersensitive reaction (HR), but one or several biochemical mechanisms may confer the inhibition of pathogen development. Several lines of evidence suggest that hypersensitive cell death *per se* does not inhibit an incompatible race, the most convincing being that certain specific metabolic inhibitors block restriction of the pathogen but not hypersensitive host cell death (see 138). The formation of induced structural or chemical barriers around the infecting pathogen is a widespread plant strategy for containment that has received substantial experimental support. Although relatively little work has been done on the role of induced structural barriers in gene-for-gene systems (but see 19), induced lignification has received strong support as a conferral mechanisms of the HR in systems involving general resistance, particularly in grasses (11,222).

Among the inducible chemical mechanisms, the release of toxicants from preformed conjugates has received substantial support in general resistance (206), but has not yet been observed in gene-for-gene systems. On the other hand, an imposing body of evidence links the production of inducibly produced phytoalexins with inhibition of pathogen development in many gene-for-gene systems (for reviews see 15, 137). The two systems for which the evidence is especially strong are the bean-*Colletotrichum lindemuthianum* system (14) and the soybean-*Phytophthora megasperma* f. sp. *glycinea* (Pmg) system (136). The status of these systems has also been bolstered by the elegant recent work of Hahn *et al.* (104) in which a radioimmune assay was used to critically study the localization of glyceollin in soybean roots after inoculation with incompatible or compatible races of Pmg. Glyceollin was detected after two hr at the pathogen infection site in plants inoculated with an incompatible race and reached an ED_{95} concentration at or before the time when cessation of fungus growth occurred; on the other hand, the phytoalexin was not detected at all in the infection site of a compatible race several hours later. In connection with the large body of evidence which has accumulated over some 15 years, the work of Hahn *et al.* makes it difficult to objectively question a role for glyceollin in the conferral of single gene resistance in the Pmg-soybean system.

Two additional significant recent papers have appeared on phytoalexins. Bell *et al.* (20) monitored messenger RNA synthesis specific for chalcone synthase, a key enzyme in green bean plants for the biosynthesis of phaseollin and other phytoalexins. They observed that the specific mRNA was markedly stimulated within a few hours after inoculation of hypocotyls with an incompatible race of *Colletotrichum lindemuthianum*. The increase occurred before the observed accumulation of phaseollin and other isoflavonoid phytoalexins and was presumed to account for their relatively large production in the incompatible hypocotyl. However, no significant increase in chalcone synthase message or phaseollin production was observed over the same time period in bean hypocotyl tissue challenged with a compatible race of the fungus. In similar work with the soybean-Phytophthora system, Schmelzer *et al.* (205) also observed rapid increases in message for phenylalanine ammonia-lyase and chalcone synthase in hypocotyls inoculated with an incompatible fungus race. The results of Bell *et al.* and Schmelzer *et al.* therefore support previous suggestions (14,234) that phytoalexin accumulation and disease resistance result from the selective stimulation in expression of a few plant genes directed to phytoalexin biosynthesis; susceptible plants fail to exhibit such changes, presumably because of lack of recognition of the compatible pathogen. It should be cautioned, however, that the relationship between messenger RNA changes and phytoalexin accumulation are as yet circumstantial. Ebel *et al.* (80) observed that xanthan gum, like a β-glucan elicitor and endopolygalacturonase, stimulated mRNA specific for chalcone synthase in soybean cells. However, glyceollin did not accumulate in response to xanthan gum, unlike the other elicitors. Thus, activation of chalcone synthase, a relatively early enzyme in the glyceollin biosynthetic pathway, does not necessarily lead to phytoalexin production. It would be of considerable interest to study mRNA levels for enzymes in the latter portions of the glyceollin pathway. Grisebach's group has provided major advances by demonstrating the production in elicitor-treated soybean cells of two specific prenyl transferase enzymes (159) and, more recently, an inducible microsomal monooxygenase (103) that stereo-specifically hydroxylates the 6a carbon of the pterocarpan nucleus to form the 6a hydroxyl group present in glycinol and the glyceollins. It would be of interest to study production of the mRNAs for these later biosynthetic enzymes after pathogen or elicitor challenge.

We do not yet understand the biochemical machinery by which primary recognition events stimulate the expression of specific host response genes such as those involved in phytoalexin production. One possibility is that the primary recognition event causes plant cell damage (hypersensitive necrosis?) such as a plasma membrane perturbation and that this trauma causes the release of 'endogenous' plant elicitors (14,69). These, in turn, elicit phytoalexin production by surrounding plant cells. The possible role of endogenous elicitors was initially suggested by observations with bean plants (13) that localized wounding resulted in the subsequent production of phytoalexins by surrounding living cells. It was later observed that pectate fragments liberated from higher plant cell walls by pectic enzymes functioned as phytoalexin elicitors (see 158,183). Inexplicably, however, soft-rotting _Erwinia_ species are copious producers of several types of chain-splitting pectic enzymes but elicit neither hypersensitive responses nor phytoalexin production in most plant tissues. Further, the radioimmune experiments of Hahn _et al_. (104) detected glyceollin in soybean roots inoculated with an incompatible race of _P. megasperma_ f. sp. _glycinea_ after only two hr.

Since this is before occurrence of hypersensitive cell necrosis, it is difficult to explain phytoalexin production on the basis of an intermediate endogenous elicitor. Also interesting was the finding by Dixon et al. (77) that denatured RNAse as well as several abiotic elicitors resulted in the release of a low molecular weight diffusible elicitor from bean suspension cells, but a biotic elicitor did not. This finding raises the question of the physiologic relevance of endogenous elicitors released by treatments such as denatured RNAse which presumably do not normally occur in the host-parasite interaction.

PROPOSED MECHANISMS CONFERRING RECOGNITION IN GENE-FOR-GENE SYSTEMS

The major current proposals are the dimer formation, the elicitor-receptor and the elicitor-suppressor models. The dimer model was formulated by Ellingboe (84) based strictly on genetic evidence. It proposed that the gene products of pathogen avirulence genes and their complementary host resistance genes physically interact to form dimers with a relatively high affinity constant. The resulting dimers were suggested to exert an unspecified inhibitory effect on development of the pathogen.

The dimer model is, in fact, a simplification of the elicitor-receptor model (7,35,137). In most versions of this latter model, agents called specific elicitors are produced by the incompatible pathogen race, the structures of which are determined by dominant avirulence alleles. The specific elicitors have high affinity for the products of their complementary plant disease resistance genes, as in the dimer model. As we have already seen, the specific recognitional event leads to, and indeed requires, subsequent selective gene expression in the plant host directed to the accumulation of phytoalexins in order to curtail development of the infecting pathogen. The dimer model predicts that only the primary gene products (proteins) of avirulence and resistance alleles interact, but most versions of the elicitor-receptor model do not preclude the possibility that secondary pathogen gene products may also be the recognitional elements. For instance, Albersheim and Anderson-Prouty (7) originally proposed that pathogen avirulence genes may code for glycosyl transferase enzymes, but that these are not the agents directly interacting with the complementary host resistance gene products. However, the unique cell surface carbohydrate structures determined by the glycosyl transferases are the agents which directly interact with the resistance gene receptor molecules. Such carbohydrates would be secondary avirulence gene products. While this version of the elicitor-receptor model has not yet been rigorously tested, it permits considerable variation in how pathogen recognition may occur (see 137). As will be seen later, there is now considerable experimental support for the occurrence of the elicitor-receptor model in several plant-parasite systems, none of which is, however, conclusive.

A major weakness of the elicitor-receptor model is that the postulated resistance gene-encoded receptors have not been unequivocally demonstrated or isolated. However, Yoshikawa _et al._ (236) recently detected specific binding sites on soybean membranes for the carbohydrate elicitor, mycolaminaran, glucans of 17-23 glucose residues, from _Phytophthora_ spp. Radiolabelled mycolaminaran specifically bound to soybean membrane preparations with a Kd value of 11.5 micromolar. Further, the binding efficiency of several mycolaminaran derivatives was closely correlated with their phytoalexin elicitor activity and mycolaminaran binding was not inhibited by several other tested carbohydrates. These results therefore suggested that the observed mycolaminaran binding site was a physiologically important receptor involved in elicitor function. It will be of interest to see whether the mycolaminaran receptor in soybean membranes is also the target for a heptaglucoside elicitor recently isolated from cell wall hydrolysates of _Phytophthora_ _megasperma_ f. sp. _glycinea_ (8). This elicitor, like mycolaminaran, is not race specific. Therefore, the mycolaminaran receptor in soybean membranes is clearly not the product of a race-specific disease resistance gene. It is appealing, however, to speculate that receptors for such widespread fungal wall components as β-glucans may comprise a general resistance mechanism which modulates phytoalexin biosynthesis in response to nonpathogens in the same way as race-specific resistance gene products. In any event, the occurrence of the mycolaminaran receptor supports the proposition that elicitors may function by interacting with specific plant protein receptors, probably located in the plasma membrane.

Pathogen molecules that suppress preformed or inducible plant defense mechanisms have been reported as a mechanism used by pathogens to overcome general resistance in plants (186,221). With gene-for-gene systems, however, the only firm evidence supporting an elicitor-suppressor mechanism has been obtained in the potato-*Phytophthora infestans* system. Doke and Kuc observed that all races of the fungus contained (and presumably display to host plant cells) non-specific elicitors, most notably the fatty acids arachidonic and eicosapentenoic acid (for review, see 144). Compatible races of the fungus have also been shown to contain mycolaminarans, which suppress the response of potato cells to the fatty acid elicitors (78). On the other hand, similar glucans from incompatible fungus races did not significantly suppress phytoalexin production. The situation has been further complicated, however, by the recent demonstration that certain fungal glucan fractions also act as 'enhancers' of the activity of the fatty acid elicitors (167). Thus, considerably more work is required to establish whether the proper mixture of elicitors, enhancers and suppressors are in fact present at the plant-pathogen interface and account for observed levels of phytoalexin production. Molecular cloning of the respective host and parasite genes governing the various gene-for-gene interactions may be required to elucidate the biochemistry of interactions in the potato-Phytophthora system. Occurrence of the elicitor-suppressor model in this system would seem to require the presence of dominant virulence genes and recessive avirulence genes in *P. infestans*, unlike other pathogens thus far investigated. However, as

Bushnell and Rowell have noted (32), biochemical models can be drawn to accommodate the elicitor-suppressor model with genetic observations.

CASE HISTORIES--SOME PROMISING ELICITORS FROM PLANT PATHOGENS

DeWit and Spikman (75) made an important discovery when they found that intercellular fluids from tomato leaves inoculated with certain races of *Cladosporium fulvum* contained race-specific elicitors of fungal origin, but that the same fungi grown on artificial culture media did not produce the metabolites. The experiment, therefore, suggested that elaboration of one or more elicitors required the plant environment. Whether this reflects a developmental effect on fungus gene expression or some other mechanism is not yet clear. DeWit *et al*. (76) have more recently characterized the race specific elicitor produced by several races of the fungus that are incompatible on cultivars of tomato containing the *Cf9* resistance gene. The elicitor activity was found to be associated with a relatively small basic peptide of molecular weight about 5000 daltons. Significantly, the peptide was not detected electrophoretically from compatible races. These most interesting findings indicate that the processed translational product of a pathogen avirulence gene may itself be a specific elicitor which is recognized by the gene product of the complementary *Cf9* tomato resistance allele. The preparation of a synthetic DNA sequence based on the amino acid sequence of the peptide would permit investigation of this question via probing a fungus genomic library for the putative avirulence gene.

The *Phytophthora megasperma* f. sp. *glycinea* (Pmg)-soybean interaction has been extensively studied and is one in which the evidence for involvement of the soybean glyceollins with resistance expression is especially strong. Unfortunately, there is still uncertainty regarding the biochemistry of specific recognition of fungus races and how these events initiate glyceollin accumulation at the infection sites of incompatible fungus races.

Ayers and co-workers (12) demonstrated that β-glucan fractions liberated from Pmg cell walls by autoclaving or acid hydrolysis were glyceollin elicitors. The same group isolated a seven-membered glucan containing β-1,6 and β-1,3 linkages that was a very active elicitor of glyceollin (8). The glucan elicitors did not exhibit race-specificity, however, and their physiologic role in plant-pathogen interactions is, therefore, unclear. As noted earlier, it is possible that they function in general resistance to non-pathogenic fungi which contain glucans in their cell walls.

Keen and Legrand (139) presented evidence indicating that glycoproteins extracted from Pmg cell walls and culture fluids gave a degree of race-specific elicitation of glyceollin in elicitor bioassays. The glycoproteins contained mannose and glucose and the carbohydrate moieties seemed to be important for elicitor activity. Yoshikawa et al. (140,235) later found that a distinct but related mannose-containing carbohydrate fraction was released from isolated fungus cell walls or living fungus hyphae by the soybean enzyme, β-1,3-endoglucanase. The released carbohydrates contained little or no protein, but were rich in mannose and also gave a degree of race-specific elicitor activity in soybean. Since the glucanase-released elicitor fraction was relatively active and would presumably be released during plant infection, it was considered to be the best candidate for a physiologically important specific elicitor. However, like the glycoprotein fractions, differences in bioassays between preparations from compatible and incompatible fungus races were only 2-4 fold, considerably less than would be expected of a physiologically important elicitor. Conclusive experiments testing the roles of the mannose-containing elicitors have not been performed, however, and recent experiments (M-C. Wang and N. Keen, unpublished) have even further clouded the picture. In a comparison of several culture media for growth of Pmg, it was found that medium composition greatly affected the structure of the mannose-containing carbohydrate fraction released from fungus cell walls by endoglucanase. The question thus remains whether the plant environment leads to production of the glucomannan elicitor fraction and whether it is physiologically important.

Ziegler and Pontzen (239) reported that the mannosylated glycoprotein enzyme, invertase, from Pmg exhibited race-specific suppressor activity. Invertase from race 3 but not race 1 suppressed activity of a non-specific glucan elicitor. However, recent attempts to fractionate invertase into protein and carbohydrate portions have resulted in complete loss of suppressor activity (E. Ziegler, personal communication). It is suspected, therefore, that the suppression observed earlier may have resulted from a non-specific interaction of invertase from only race 3 with the glucan elicitor. As with the potato-Phytophthora system, cloning of the relevant host resistance and pathogen avirulence or virulence genes may be required to interpret the confusing biochemical work in the Pmg-soybean system.

Anderson and co-workers (9) observed that culture fluids of *Colletotrichum lindemuthianum* contained race-specific factors which elicited phytoalexin production in incompatible but not compatible green bean cultivars. The elicitor from the α-race has been purified and the specific elicitor activity was associated with a glycoprotein of molecular weight ca. 62 kD, which at least partially degraded into two smaller molecules (220). Significantly, the glycoprotein was not observed in similar preparations from two other compatible fungus races, and no evidence was obtained for the occurrence of race-specific suppressors by any fungus race. Thus, as in the *Cladosporium*-tomato system, it is possible that a protein elicitor may be functional in the *Colletotrichum*-bean system but it has not yet been established whether the activity is due to the carbohydrate moiety. It should also be possible to ask whether the α-race specific glycoprotein elicitor is the primary transcript of a *Colletotrichum* avirulence gene.

Mayama and Tani and their co-workers obtained evidence strongly suggesting that the production of phytoalexins, called avenalumins, is the basis for expression of single gene resistance in oat plants to incompatible crown rust races (171,218). One of the genes investigated, *Pc-2*, is well known because it conditions sensitivity to victorin, a host-specific toxin produced by *Helminthosporium victoriae* (for review see 62). Mayama and Tani (173) observed that partially purified victorin preparations as well as inoculation of plants with *H. victoriae* resulted in substantial production of the avenalumins in *Pc-2* but not other resistance genotypes of oats. They also made the important observation that, unlike the crown rust fungus, the growth of *H. victoriae* is not severely inhibited *in vitro* by victorin. Victorin has recently been purified and found to be a series of related, low molecular weight peptides (142,224,228). The elicitor experiments have recently been confirmed using near-isogenic oat lines with and without the *Pc-2* allele and a highly purified victorin preparation (172). Again, *Pc-2* oat leaves accumulated the avenalumins in response to victorin concentrations as low as 1-5 pg ml^{-1}, but *pc-2* oats were unaffected at 100 ng ml^{-1}. Significant plant cell death did not occur at low victorin concentrations, but higher concentrations (1 ng ml^{-1} or higher) resulted in substantial necrosis and reduced phytoalexin production by *Pc-2* leaves. All of these observations therefore support the early suggestion of Litzenberger (163) that victorin in fact initiates a massive hypersensitive reaction in *Pc-2* oats, but that *H. victoriae* continues growth in the affected oat tissue. Victorin, therefore, appears to mimic an avirulence gene product of the crown rust fungus which is complementary to a product of the *Pc-2* allele. Thus, victorin constitutes a probe with which to attempt isolation of the putative receptor protein encoded by the *Pc-2* allele. Since victorin has been purified and structure determination of the active peptides is in progress by Dr. V. Macko of the Boyce Thompson Institute, it is reasonable to anticipate that an *in vitro* labelled probe can be synthesized from victorin in order to search for the putative receptor protein of *Pc-2* oats.

Production of the race-specific elicitors

discussed above probably will be experimentally related to the putative pathogen avirulence genes that produce them in the near future. Recent molecular cloning experiments with bacterial pathogens have developed along the opposite line--avirulence genes have now been cloned from two different bacterial pathogens and the search in these cases is for the identity of their primary gene products and investigation of whether the primary or secondary gene products function as specific elicitors.

THE MOLECULAR CLONING OF PATHOGEN AVIRULENCE GENES

Staskawicz *et al.* (215) first cloned an avirulence gene from *Pseudomonas syringae* pv. *glycinea* (Psg). This plant-pathogen system had previously been shown to behave as a gene-for-gene system by physiologic experiments and limited crossing of host cultivars (138). In all studied race-cultivar combinations, a strict correlation between glyceollin accumulation and bacterial multiplication has been observed (141,164). It is not yet firmly established, however, whether the glyceollins are in fact responsible for the bacteriostasis observed in hypersensitive resistant reactions (138). Bruegger and Keen (27) detected race-specific elicitor activity in cell envelope fractions from four races of the bacterium, but were unable to isolate the elicitors. In order to more systematically approach gene function in the soybean-Psg system, an attempt was made to molecularly clone avirulence genes from various bacterial races. A cosmid library of race 6 DNA was constructed in pLAFR-1 and one clone was isolated which conferred race 6 incompatibilities when transferred to other Psg races by conjugation (215). Mutagenesis with Tn5 and sub-cloning were employed to obtain a ca. 3.0 kb Acc I fragment from the initial cosmid clone which conferred the race 6 phenotype when introduced into race 4 Psg on the plasmid pRK404. The cloned avirulence gene has not yet been translated into protein, but attempts are underway to isolate the avirulence gene product and assess its phytoalexin elicitor activity.

Significantly, the sub-cloned race 6 avirulence gene was also expressed when introduced into *Xanthomonas campestris* pv. *glycines*. The tested *Xanthomonas* isolate was pathogenic on all of the standard soybean differential cultivars for Psg, but introduction of the race 6 Psg avirulence gene resulted in precisely race 6 Psg host reactions by the soybean differentials--that is, a hypersensitive reaction was observed on four cultivars which are incompatible to Psg race 6, but not on three compatible cultivars (B. Staskawicz and D. Dahlbeck, unpublished). These results would argue against a cell surface carbohydrate being the specific elicitor of glyceollin production in the Psg-soybean system, since it is difficult to envision the primer carbohydrates of *Xanthomonas* being utilized by a Psg glycosyl transferase. The availability of the cloned race 6 avirulence gene, however, permitted a definitive test of whether the lipopolysaccharides of Psg function as race specific elicitors.

Barton-Willis *et al.* (pers. comm.) characterized the structure of lipopolysaccharides (LPS) from four races of Psg. Since the O-chain carbohydrates typically exhibit the greatest variation in structure (138), these were characterized in detail using methylation analysis, Smith degradations and ^{13}C and one- and two-dimensional ^{1}H NMR data. The O-chain structures of races 1, 5, and 6 were indistinguishable (Barton-Willis *et al.*, in preparation). The repeating unit of these LPS contained a rhamnose backbone with a rhamnose side chain terminated by an N-acetyl glucosamine residue. Race 4 had a much larger repeating unit but it was established that the difference in O-chain structure between race 4 and races 1, 5, and 6 was not related to race phenotype. This was deduced by conjugating the cloned race 6 avirulence gene into race 4 and studying the O-chain carbohydrates from the resulting transconjugant. No difference was detected in the LPS O-chain structure from the race 4 transconjugant, but it behaved phenotypically like race 6. This observation, coupled with the lack of difference between the O-chain structures of races 1, 5, and 6, established that they could not be the race specific elicitors of Psg.

Gabriel (95) recently reported the molecular cloning of five different avirulence genes from *Xanthomonas campestris* pv. *malvacearum* which complemented five different resistance genes in cotton. Transfer of the cloned avirulence genes to a compatible bacterial race resulted in the expected hypersensitive reactions only on cotton lines carrying the complementary disease resistance genes. Sub-cloning and characterization of the Xanthomonas avirulence genes should yield considerable information on the molecular biology of the host-parasite interaction.

Finally, R. Stall (214) has recently demonstrated that race 1 of *Xanthomonas campestris* pv. *vesicatoria* carries an avirulence gene on an indigenous plasmid that also codes for copper resistance. Since the plasmid can be mobilized, Stall introduced it into other *Xanthomonas campestris* strains by conjugation and observed that the avirulence phenotype co-inherited with copper resistance. The avirulence gene has not yet been cloned and characterized.

CONCLUSIONS AND FUTURE CONSIDERATIONS

The recent isolation of race-specific elicitors, particularly from *Colletotrichum lindemuthianum* and *Cladosporium fulvum*, along with the recent cloning of bacterial avirulence genes suggests that the primary protein products of avirulence genes may function as specific recognitional elements in these systems. While it is probable that much greater complexity will be observed as our knowledge increases, including cases of suppressor function, the biochemical and molecular genetic evidence thus far are fully in accord with the classical observation that avirulence genes are inherited as dominant genetic characters and probably dictate specificity (84).

Additional research will result in the further characterization of some of the promising examples of race-specific pathogen elicitors noted in the earlier sections. The tools of recombinant DNA technology should permit an understanding of the relationship between these elicitors and pathogen avirulence genes--are the elicitors primary or secondary avirulence gene products? How do suppressors relate to the function of virulence/avirulence alleles? Are plant resistance gene products the protein receptors for pathogen-produced specific elicitors as the elicitor-receptor model predicts? If so, how does the elicitor-receptor complex initiate the observed selective transcription of plant genes involved in phytoalexin biosynthesis? Answers to the first question are imminent since translation of the avirulence genes cloned from bacterial pathogens will permit elicitor bioassays to be performed with these proteins. The peptide elicitors isolated from *Cladosporium fulvum* and *Colletotrichum lindemuthianum* permit the screening of genomic libraries of the respective fungi in order to isolate the respective avirulence genes. The availability of these specific elicitors as well as victorin and others such as the protein products of cloned bacterial avirulence genes will be used as probes to search for the as yet hypothetical resistance gene-encoded peptides. If these peptides are isolated, they in turn permit a direct approach to the molecular cloning of the plant disease resistance genes from genomic libraries. The availability of characterized avirulence and resistance gene products will also permit investigation of the molecular events occurring between initial pathogen recognition and host gene activation--are endogenous elicitors produced by the primary recognitional event or is the elicitor-receptor complex transported to the plant cell nucleus or are second messengers involved? Finally, all of the above research will decisively test the physiologic role of phytoalexins in resistance expression and perhaps suggest ways in which we can modulate the phytoalexin responses of higher plants for our benefit.

One of the anticipated returns from studies on the molecular biology of plant-pathogen interactions is the possibility of transferring disease resistance genes between plants that cannot be intercrossed. However, it is unlikely that single genes conferring race-specific disease resistance will be particularly useful in this regard because they are too specific--for example, a resistance gene for crown rust resistance in oats indeed has utility in other oat cultivars, but this exchange can easily be accomplished by conventional plant breeding. On the other hand, the crown rust resistance gene is unlikely to confer resistance to *Phytophthora* root rot in soybean if the gene could be cloned and introduced into that plant by molecular genetic techniques. A more appealing rationale would be to isolate a plant gene which confers general (non race-specific) resistance to an entire pathogen species, pathovar or *forma specialis* and introduce it into a plant that is susceptible to the pathogen. For instance, the various *formae speciales* of rust and mildew fungi and pathovars of *Pseudomonas syringae* and *Xanthomonas campestris* are usually pathogenic on only one plant species and elicit hypersensitive defense reactions on other plant species. Single genes for such general resistance have not often been detected because relatively few plants of different genera can be intercrossed. In the cereals, however, intercrossing and/or the use of gene translocations has allowed disease resistance genes to be incorporated from relatively disparate grasses. Some of these genes, such as *Sr25* in wheat (175), have proven to be remarkably resilient to the emergence of new virulent pathogen races, as frequently occurs when race-specific resistance genes are used. This experience, coupled with the recent ability to clone race-specific pathogen avirulence genes, encourages the thought that general or non race-specific avirulence genes can be isolated from fungal *formae speciales* or bacterial pathovars which mediate non-pathogenicity on a certain non-host plant(s). Cloning of the corresponding complementary general resistance gene from such a non-host plant would then permit, via a Ti plasmid vector, its introduction into an unrelated plant that is normally a host for the pathogen. It is appealing to predict that the recipient plant would exhibit non-pathogen general resistance to the entire pathovar or *forma specialis* of pathogen. Furthermore, since such resistance would not be race-specific, the evolution of the pathogen to new, virulent races should be impeded.

ACKNOWLEDGMENT

The author's research is supported by grants from the National Science Foundation.

PROSPECTS FOR USING RECOMBINANT DNA TECHNOLOGY TO STUDY RACE-SPECIFIC INTERACTIONS BETWEEN HOST AND PARASITE

Albert H. Ellingboe

Professor of Plant Pathology and Genetics
University of Wisconsin, Madison, Wisconsin 53706

The perception of how molecular genetics, and recombinant DNA technology in particular, has affected and will affect plant pathology is varied. Many plant pathologists have indicated that recombinant DNA technology is only a passing fad, to be ignored because it will have no significant effect on plant pathology, and that plant pathologists need to get on with the work that plant pathologists do. If I were to agree with this view, this paper could stop here. But I do not agree with this view because it is clear to me that the arguments of molecular genetics and recombinant DNA technology are already having an impact on plant pathology. For example, plant pathology is already being influenced by recombinant DNA technology in the form of diagnostic kits for viruses. There are large numbers of genetically engineered biocontrol agents ready to be tested as soon as approval to test is granted. Several avirulence genes from prokaryotic pathogens have been cloned. Considerable progress is being made toward developing routine genetic transformation in plants.

Probably the most significant effect of molecular genetics and recombinant DNA technology is the change in thinking about biological phenomena. The way in which questions are framed, the kinds of questions asked, and the procedures have dramatically changed in the past decade. The ability to use the arguments of molecular genetics in studies of host-parasite interactions is dependent on an accurate description and interpretation of the basic genetics of the interaction. This article is an attempt to put the basic genetics of interactions in a perspective that permits an analysis using the new technologies applied to the genes controlling those interactions. The goal is to clone those genes and determine their products. Subsequently, there will be numerous uses of the clones and the gene products that will greatly aid answering other questions about how a plant or plant part is resistant or susceptible and how a pathogen is avirulent or virulent.

DEMONSTRATION OF VARIABILITY

After the rediscovery of Mendel's laws it was obvious that there was an opportunity to use these concepts to modify plants through breeding. Observations had been made earlier that there were large differences between different seed stocks within a single plant species in how they react to disease. The reaction to disease was shown to be inherited--usually in a relatively simple manner--and then was used extensively in breeding disease resistant cultivars. The development of disease resistant cultivars led to the concept of races of the pathogens. Race recognition was a way to account for the observation that a cultivar that was resistant in one year and/or geographic area could, in subsequent years or in different areas, be susceptible. The cultivar did not change; the pathogen changed. The interrelationships between the host genetics and the pathogen variability was not understood until the mid 1940's. Flor (89,90) made genetic analyses of the interactions by making a genetic analysis in both host and parasite. What he found was a very specific interdependence of the genetic systems in both host and pathogen. A plant could not be resistant unless the pathogen had the corresponding gene for avirulence. A pathogen could not be avirulent on a plant unless the plant had a corresponding gene for resistance. Though many host genes and pathogen genes were observed to segregate in crosses of each species, each host gene had only one corresponding pathogen gene (and vice versa) for resistance and avirulence, respectively (Table 1). Flor called this pattern of interaction a gene-for-gene relationship. The pattern is illustrated in its simplest form in Table 2.

Table 1. The pattern observed by parents and progeny in crosses of both host and pathogen in which differences in compatibility occur at only one locus in the host and one locus in the pathogen.

		Host			
		Parents		Progeny	
		A	B	F1	F2
Pathogen strain	Genotype	R1R1	r1r1	R1r1	R1-:r1r1
Parents					
1	P1P1	-	+	-	3-:1+
2	p1p1	+	+	+	all +
Progeny					
F1	P1p1	-	+		
F2	P1-	3-	All +		
	p1p1	1+	All +		

\- = incompatible interaction (host resistant, pathogen avirulent).

\+ = compatible interaction (host susceptible, pathogen virulent).

Table 2. The basic pattern of compatibility in two host lines and two pathogen strains which shows the interdependence of host and pathogen genes.

Pathogen genotype	Host genotype	
	R1R1	r1r1
P1P1	-	+
p1p1	+	+

- = incompatible interaction (host resistant, pathogen avirulent).

+ = compatible interaction (host susceptible, pathogen virulent).

UNIVERSALITY OF GENE-FOR-GENE SYSTEM

Shortly after Flor's publication, I believe it is fair to say that most researchers considered the gene-for-gene interactions to represent a very unique, specialized system of interactions restricted to very specialized types of parasites. It is now known that this is not true. The gene-for-gene pattern of interactions has been demonstrated, or shown to be consistent with, the data on host-parasite systems for almost all systems investigated. I believe it is fair to say that more than 95% of the within-species variability follows the gene-for-gene pattern. That includes whether the differences between + and -, as given in Figs. 1 and 2, are large or small, differences are in infection type, infection efficiency, growth rate of the pathogen, rate of development of an epidemic, or whatever difference was observed. The essence of the pattern is present, namely, the interdependence of the genes in both host and parasite and, upon detailed analysis, the one gene-for-one gene pattern controlling the specificity of the interactions. There have been numerous occasions where the differences have appeared to be other than in a gene-for-gene type of interaction but upon more detailed analysis the gene-for-gene pattern has emerged. Almost all genetic variability in disease compatibility manipulated in plant breeding follows the gene-for-gene pattern. From the standpoint of what is of commercial interest, it is necessary to understand the role that these genes play in practical disease control by breeding, and to put into proper biological perspective any efforts to understand molecular mechanisms by which the host and pathogen interact.

TYPES OF GENES POSTULATED TO CONTROL INTERACTIONS BETWEEN HOST AND PARASITE

The gene-for-gene pattern (See Table 2) can best be explained by assuming that the primary product of an R gene interacts with the primary product of a P gene to give an incompatible interaction. A specific interaction gives incompatibility (host resistant, pathogen avirulent). A lack of interaction on the part of the corresponding gene products gives compatibility (host susceptible, pathogen virulent). Resistance is due to production of a particular product. Avirulence is due to production of a particular product. Resistance is an active function. Avirulence is an active function. Almost all naturally-occurring genetic variability affecting host-pathogen interactions follows a pattern of active function on the part of the microbe to be unable to grow in the host. Therefore, the gene-for-gene interactions were assumed to be superimposed on a basic compatibility between host and pathogen (82).

It seems reasonable to postulate that there are genes in a pathogen that carry on an active function to penetrate into and grow in tissues of the host. Why variability for these genes is not observed in nature is an important question that will not be dealt with here. Instead, an examination of the evidence for such genes will be given. Basically three types of evidence are available.

The research on the interactions between *Rhizobium* and its host plant began with the assumption that there are genes in the bacterium that have a positive function for nodulation and nitrogen fixation. Mutants were induced that lacked the ability to nodulate or fix nitrogen (165). Members of a library from a nod^+ fix^+ strain were transformed into the mutant culture, and transformants with the ability to nodulate and/or fix nitrogen were selected. If the member of the library that has the wild type allele for the mutant gene carries on an active function, then the resultant merodiploid should nodulate and/or fix nitrogen. By this procedure (basically cloning by function) it was possible to show that there are genes in the microbe that do carry on an active function for the ability to grow in the host plant (165).

A second line of evidence for active function for virulence comes from the work with host-specific toxins (196). There are molecules produced by microbes that affect the ability of the microbes to grow in plants of a given species or of a given genotype of a species. *Helminthosporium victoriae* produces a toxin that permits the fungus to grow in *Avena byzantina* plants if the plants have a dominant *Vb* gene. The results suggest not only that *H. victoriae* produces a particular molecule, but also that the host plant, to be susceptible, must produce a specific receptor molecule. The results with toxins are based on molecules that can be bioassayed in some manner.

A third line of evidence for active function for virulence comes from the research on temperature-sensitive mutations in pathogens (85). Most temperature-sensitive mutations affect the tertiary structure of the protein gene product. An amino acid substitution can lead to a functional gene product at a normal temperature but a nonfunctional product at a high temperature because of the loss of tertiary structure. In a pathogen, a temperature-sensitive mutation in a gene whose function is necessary for infection and growth in the host should be able to infect and produce disease at the normal temperature but not at the higher temperature (a temperature at which the wild type can infect and produce disease). Temperature-sensitive mutations have been produced that do not give temperature-sensitivity as measured by growth on agar media (complete or minimal) but are temperature-sensitive in the host in that they permit growth of the pathogen, and disease, at the normal temperature but not at the higher temperature (85).

The evidence for the existence of genes in pathogens that carry on an active function for infection and growth in the host is based on three types of experiments, namely, complementation of induced noninfective mutations (essentially cloning by function), bioassay for toxic metabolites that induce similar symptoms as the pathogen and/or promote growth of the pathogen in the host, and the genetic arguments associated with induced conditional mutations (primarily temperature-sensitive mutants that do not permit pathogen growth at the elevated temperature). Why is more genetic variability not seen within a pathogen species based on genes that have a positive function for growth in the host? Almost all naturally-occurring genetic variability in interactions between a pathogen species and a host species analyzed to date follows the pattern of a gene-for-gene relationship, an interaction in which avirulence is an active function. Is variability of genes whose functions are crucial for successful infection and growth in the host obviated by selection? Are such mutants deleterious in an evolutionary sense?

It is the naturally-occurring genetic variability that is manipulated in commercial plant breeding. I am primarily interested in genes that have agronomic value. That is the reason why, in the remainder of this discussion, primary consideration will be given to those genes for which there are naturally-occurring variants.

Much has been learned from the formal Mendelian genetics of the naturally-occurring genetic variability in host-parasite interactions. The numbers of genes gives an indication of the complexity of the interactions. Wheat, for example, has at least 34 Sr loci for reaction to Puccinia graminis, 40 Lr loci for reaction to Puccinia recondita, 8 Pm loci for reaction to Erysiphe graminis, named thus far. Though 34 loci may certainly suggest that resistance to P. graminis is controlled by many genes, it is possible to select host lines and pathogen strains so that each locus can be monitored independently.

Host R genes are usually in allelic (or pseudoallelic?) series (90). Pathogen P genes are usually not in allelic series. There are alleles at several Sr loci so that the total number of R genes is greater than 34 and the number of corresponding P genes is also greater than 34. Some host allelic series appear to be true alleles (L locus in flax) whereas other host allelic series appear to be closely linked R loci (pseudoalleles, e.g. M locus in flax and Rpl locus in maize) (209).

Mutational analyses of P and R genes have shown that it is relatively easy to get mutations in specific genes in plants from resistance to susceptibility and from avirulence to virulence in pathogens on host lines with specific R genes. Mutations to resistance at known R loci have not been observed (to my knowledge), and mutations in pathogens toward avirulence to host plants with specific R genes have been found only very rarely.

As already noted, the temperature-sensitive interactions of naturally-occurring genetic variability in either host or pathogen are usually of the pattern whereby the plant is resistant (pathogen avirulent) at a normal temperature and more susceptible (pathogen virulent) at a higher temperature. These data are consistent with the interpretation that the specific interactions are for incompatibility (host resistant, pathogen avirulent).

Classical genetic analyses have almost always shown a clear one gene in the pathogen for one gene in the host relationship. The strict adherence of the one-for-one relationship for almost all naturally-occurring genetic variability suggests that the crucial interactions are between primary products of the P and R genes (82,84). If the crucial interactions were between secondary products, deviations from the one-to-one relationship might be expected. A popular hypothesis in the past few years is that there are two kinds of genes (regulator and regulated) (84). One set of genes that follow the gene-for-gene relationship are regulatory in nature. They regulate genes that make the molecules that make a plant resistant or susceptible and/or a pathogen avirulent or virulent. If this hypothesis were correct, variants of the regulated genes would be expected as well as of the regulator genes. Such variants would appear as deviations from the strict one-to-one relationship that is consistently observed in experiments on inheritance of variability in interactions. The regularity with which the one gene-to-one gene pattern is observed has led me to conclude that it is the primary gene products that are crucial in deciding whether a host-parasite interaction will be compatible or incompatible. Therefore, it is the primary gene product that is of greatest interest to me.

GENETIC DESCRIPTION OF A GENE

The differences between different host lines and pathogen strains discussed above have been determined by crossing via the standard sexual cycle and classical genetic analysis. In such crosses recombination occurs between the genomes from each parent. The entire genome participates in the recombinational process. The segregation of phenotypic differences among progeny of these crosses represent the data from which the interpretations were made.

In classical Mendelian genetics, a gene is identified only if a variant of that gene exists. Without a variant to observe segregation in progeny, the gene would remain unknown. The segregation patterns can be observed for several genes to see if any tend to segregate together and thus establish a linkage map. If the individual chromosomes are identifiable by one or more of a number of different techniques, it may be possible to assign a gene and a linkage group to a particular chromosome or chromosome segment. The temptation is to conclude that the gene which is known to control a particular phenotypic effect is associated with a chromosomal segment, or a segment of DNA, while in fact the reverse is true. A segment of DNA is associated with the theoretical construct of the gene that is known to control a particular phenotypic effect. This is not a trivial distinction because it relates to the arguments used in classical and molecular genetics.

COMPARATIVE STUDIES OF INTERACTIONS

There has been both extensive and intensive research over the past four decades to understand the mechanisms of resistance to pathogens in plants, and the mechanisms of pathogenicity and virulence in pathogens of plants. The studies used several procedures. The comparative biochemistry and physiology has been examined between inoculated and noninoculated plants, between inoculated plants where the resistant plant was one species and the susceptible plant was a different species, between inoculated resistant and susceptible plants of the same species, between highly isogenic lines of the same species, and between progeny in segregating generations of crosses of resistant and susceptible plants. These studies have shown many molecular changes associated with resistance or susceptibility, but they have not led to an identification of the products of the *R* genes. Similar types of experiments have been made with pathogens but again the effort to identify the primary products of the *P* genes has been unsuccessful. There appear to be so many physiological and biochemical changes during the expression of interactions between a plant and a pathogen that it has been very difficult to distinguish which changes are required for a particular interaction and which are the result of a particular interaction. The genetics of the interactions are quite simple, but the consequences of those genetic differences appear to be exceedingly difficult to sort out in terms of cause and effect relationships.

RECOMBINANT DNA TECHNOLOGY IN CROWN GALL RESEARCH

The advent of the recombinant DNA technology in molecular genetics in the past decade has permitted a series of different approaches to biological studies of many kinds. Plant pathology has not escaped the impact of this change in viewing biological phenomena. In fact, it has played a rather central role in the application of the arguments of molecular genetics and the recombinant DNA technology to plants. The disease that has made this possible is crown gall caused by *Agrobacterium tumefaciens*. Galls that formed that were freed of bacteria were found to contain DNA that was not present in the nontransformed plant but was present in the bacterium prior to inoculation into the plant (49). The unique piece of DNA was found to be on a plasmid whose presence was necessary for the formation of a gall. One segment of the plasmid, now called the T DNA, was shown to be unique in that it was transferred from the bacterium to the host genome. The initial suggestion that DNA of bacterial origin was transferred from the bacterium to the host plant was based on reannealling kinetics of total DNA from bacterium and host plant. The DNA transfer was subsequently confirmed by demonstrating that a unique segment was present in the host DNA by cloning the segments of the Ti plasmid and using each segment as a probe for DNA from transformed and nontransformed plants. The probes made from segments of the Ti plasmid have given clear, qualitative evidence for the existence of T DNA in transformed plants. The T DNA has been shown to contain several transcripts, and the role of several transcripts in gall morphology and biochemistry has been demonstrated (71). Parts of the Ti plasmid DNA other than the T region have been shown to play roles in infection and physiology of the gall produced.

The Ti plasmid codes for functions that are necessary for gall formation. There is evidence for active gene function for gall formation in the crown gall disease. There appears to be little or no within host species genetic variation in reaction to *A. tumefaciens*. On these two observations, the data on crown gall disease suggest that the genes involved in gall development are not the same kinds of genes that are involved in race specificity. The pattern is different. This does not mean the research is unimportant. It means that the research will likely not bear on the question of the basis for race specificity. It does clearly show that the arguments and techniques of molecular genetics can be brought to bear on questions of host-parasite interactions, and that definitive answers have been obtained where only vague correlations were observed in other experimental approaches.

CLONING OF *P* GENES

There are volumes of literature on the molecular changes associated with the expression of resistance and susceptibility. The problem has been to determine which of the myriad of molecular differences is a key to determining the specificity of the interaction. The advent of the recombinant DNA technology has changed the perspective of the problem. Rather than start from a population of molecules and work to sort out which is important, it is now possible to start with the DNA segment that is known to control the phenomenon and work from the gene toward the phenotype. The segment of DNA is recognized in a library of DNA by its demonstrated effect on a phenotype, the same procedure used to identify the gene in the first place (i.e., the effect of a gene variant on the phenotype). The procedure now becomes one of first cloning the DNA segment that contains the gene and then moving toward the phenotype. Two illustrations as to how these procedures are being used to identify *P* and *R* genes will be given in the following paragraphs.

Pseudomonas syringae pv. *glycinea* is the agent that incites bacterial blight of soybeans. Soybean cultivars differ in their reaction to different strains of the pathogen. Some soybean cultivars give a hypersensitive reaction when inoculated with a particular strain of the pathogen. This reaction is considered to be an incompatible interaction, that is, host is resistant and pathogen is avirulent. Other host cultivars may give a compatible reaction to the same strain, and the cultivar would be considered to be susceptible and the pathogen considered to be virulent. The reaction of four cultivars to four races of the pathogen is given in Table 3. Also given are the postulated genotypes of each race and each cultivar. The postulated genotypes are based on the assumption that the gene-for-gene relationship holds for this parasite/host system. The three parasite/host gene pairs postulated will explain the data in Table 3. The incompatible reactions are assumed to

Table 3. The reaction of four soybean cultivars to inoculation with four races of _P. syringae_ pv. _glycinea_, and the postulated genotypes of host cultivars and pathogen races.

	Soybean Cultivar and Postulated Genotype			
	Flambeau	Harasoy	Norchief	Peking
Pathogen race	r1 R2 r3	R1 r2 R3	r1 R2 R3	R1 r2 r3
1 p1p2P3	+ [a]	-(3) [b]	-(3)	+
4 p1p2p3	+	+	+	+
5 p1P2p3	-(2)	+	-(2)	+
6 P1p2p3	+	-(1)	+	-(1)

[a] + = compatible reaction
- = incompatible reaction (hypersensitive reaction)

[b] = The gene pair that gives incompatibility is given in parentheses.

Data from Staskawicz, _et al_. (209).

result from the specific interactions of corresponding P and R genes. The interpretation given, based on the gene-for-gene model, is that, for example, the product of P1 in race 6 interacts with the product of R1 in Peking to give an incompatible interaction. If the gene P1 in race 6 were to be transferred to race 5, and if the above interpretation were correct, then the merodiploid (race 5 + DNA that contains P1 from race 6) would be expected to be avirulent on Peking. If race 6 was avirulent on Peking because it lacked something that race 5 had, then the merodiploid would be expected to remain virulent on Peking. The hypotheses were tested as follows. A genomic library of race 6 P. syringae pv. glycinea was constructed in the cosmid vector pLAFR1 (215). The average size of the insert was about 25 kilobases. That means that the complete genome of race 6 should be present in a library of about 600 clones (tests on a known single copy gene also indicated that 600 clones would represent the complete genome). The library was transconjugated into race 5. Six hundred and eighty cultures were isolated and tested by inoculation into the four cultivars listed in Table 1. One culture, with clone designated pPg6L3, had an altered reaction on cultivars Peking and Harasoy. None of the cultures had reactions on Flambeau and Norchief cultivars different from race 5. One DNA fragment 27.7 kilobases in length contained the ability to convert race 5, which is virulent on Peking and Harosoy, to be avirulent on Peking and Harosoy. These results are consistent with the hypothesis that avirulence is a positive function, and that, in keeping with the interpretation of the Mendelian genetics, the specific interactions are for incompatibility. None of the transconjugants had been converted from avirulent to virulent.

At least one gene controlling race specificity is on a 27.7 kb fragment of DNA. The amount of DNA of interest for specificity on Peking has been reduced by a factor of 600. But 27.7 kb of DNA can contain several genes. The location of the gene(s) determining specificity on the 27.7 kb segment was determined by transposon mutagenesis and digestion with restriction endonucleases. Transposons are segments of DNA that move about in the genome. When a transposon moves from one location in the genome to another location, it will usually inactivate the gene into which it inserts. Insertion into a particular gene is monitored by a change in phenotype. The transposon Tn5 was used to mutagenize the cosmid clone containing the 27.7 kb DNA fragment. Of 180 independent insertions of Tn5 into the cosmid, four no longer had the ability to convert race 5 from virulent to avirulent on cultivar Peking. A combination of mapping by digesting with restriction endonucleases and Tn5 mutagenesis helped to locate the region coding for race-specificity to approximately one tenth of the original 27.7 kb fragment. The segments containing the ability to convert race 5 from virulence to avirulence (the *P* gene) on Peking could now be subcloned.

There are several possible ways to determine the product of the *P* gene on this fragment. One is to sequence the DNA to determine the open reading frames. If the start and stop codons could be identified, a knowledge of codon preference would permit a postulated primary structure of the protein. From a knowledge of the primary structure, the size could be estimated, the hydrophobicity of segments could be deduced, the possible site of the protein in the cell could be deduced, antibodies could be produced against synthetic polypeptides that mimic the postulated primary structure and the antibodies used to find the native primary gene product, etc. A sequence alone will open several avenues of investigation of determining *P* gene products in pathogens.

A second strategy to determine the product of a *P* gene is to put the gene on a high expression vector. Can *E. coli*, yeast, *Xenopus oocytes*, or some other organism be made to make sufficient gene product so that isolation of the product would be relatively easy? Other strategies are available, and new ones will undoubtedly continue to be discovered.

At least three parasite/host gene pairs are necessary to explain the differential interactions given in Table 3. Three is the minimum, not the maximum. If the postulated genotypes are correct, predictions can be made of segregation patterns in crosses of the host lines. Further analyses of libraries of the four races should show no additional _P_ gene from race 6, one _P_ gene from race 5, one _P_ gene from race 1, and no _P_ gene from race 4 controlling interactions with these four cultivars, if the postulated genotypes are correct.

The above procedures for cloning avirulence genes have been used with other bacterial pathogens. Seven different clones of _Xanthomonas campestris_ pv. _malvacearum_ have been obtained which show different spectra of interaction with each of a set of highly isogenic host lines (95). Five clones, each distinct, convert a virulent isolate to an avirulent isolate on one of five different host lines each with a single and different _R_ gene. So the results with _P. syringae_ pv. _glycinea_ are not just a peculiarity of one clone. The procedure of cloning for _P_ genes seems applicable to other pathogens as well. Once the genes are cloned, there should be no insurmountable problems in determining the products of the _P_ genes. The technology is available.

Are there homologies between _P_ genes? If there are, will cloning of one _P_ gene make probing for other _P_ genes easier? The initial studies on homology between different races with the _P_ gene cloned for _P. syringae_ pv. _glycinea_ were surprising (215). The clones of two segments (each identified as being part of the _P_ gene by transposon mutagenesis) apparently had no homologous regions in the three other races analyzed! Does that mean that the locus is missing in the other races? Can segments of partial homology (the alternate alleles) be identified if the stringency of the hybridization conditions are reduced? These are nagging questions that will be answered as more _P_ genes are cloned and tested against alternate alleles and clones of _P_ genes with different specificities.

R GENES IN HOST PLANTS

The research on cloning _P_ genes in pathogens was facilitated by the availability of genetic transformation in the bacterial pathogens. The ability to do routine transformation in plants is much more restricted.

Efforts are also being made to clone _R_ genes from plants. The procedures outlined briefly above for the bacterial pathogens are not acceptable for most plants for several reasons. Plants have considerably more DNA than bacteria. If individual clones are 20-30 kb, it will take 200,000 to 500,000 clones to have reasonable assurance of having a complete library. It is now possible to do transformations of dicotyledonous plants with the Ti plasmid from _Agrobacterium_. The thought of having to redifferentiate and screen 200,000 to 500,000 plants to get a single genetic transformat is horrendous. There are ways to reduce this number if the screen for resistance were to accept any of a number of genes simultaneously. For example, if transformation of any of ten different _R_ genes in the library would give resistance to the race of the pathogen used, the number of transformed plants to be tested could be reduced by a factor of ten.

For most plant species, a system for routine transformation is not available. Though considerable effort is being expended to find one, a genetic transformation system for maize, wheat, rice, barley, oats, and other grasses is not presently available. Therefore, a different approach has been used to identify _R_ genes in maize. The approach is to use transposable elements to insertionally inactivate _R_ genes. If the transposon is cloned, the clone of the transposon can be used as a molecular probe to identify and clone the flanking segments of the _R_ gene, which can then be cloned to identify the nonmutated _R_ gene. In the following paragraphs an example will be given of a procedure being used by the author and cooperators to attempt to clone the _Rp1f_ gene for resistance to _Puccinia sorghi_ in maize.

The first step was to get a clone of a plant transposable element. This was accomplished by comparing a wild type alcohol dehydrogenase gene with a suspected Robertson's Mutator (Mu1)-induced mutation in the alcohol dehydrogenase gene (22). The genomic clone of wild type was compared to a genomic clone of a Mu1-induced mutation of the alcohol dehydrogenase gene. The difference between the two clones was a 1.4 kb segment of DNA in the first intron of the gene. The DNA in this insertional sequence was cloned and sequenced. The sequence was shown to have the characteristics of transposable elements, such as terminal repeats, and to be present only in maize lines that had been crossed with, or derivatives of, plants that possessed mutator activity. Plants that had no history of crosses with plants with Robertson's Mutator activity had no sequences homologous to the clone of Mu1. The conclusion was that the mutation of the alcohol dehydrogenase gene was caused by the insertion of the 1.4 kb DNA fragment into the first intron of the gene.

The crossing scheme used to put Mu1 into the Rp1f gene is given in Fig. 1. Plants homozygous for Rp1f were crossed with rust susceptible plants known to have mutator activity. The F1 plants were selfed. Progeny that were homozygous for Rp1f were crossed with plants of genotype rp1/rp1 oy/oy. The oil yellow gene, oy, is linked to rp1. The heterozygous progeny Rp1f/rp1 Mu1 Oy/oy were inoculated with *P. sorghi*. All plants should be expected to be resistant to the isolate used. Thirty-eight susceptible plants were obtained following inoculation of over 35,000 plants. Progeny of 29 plants were obtained either by selfing or test crossing. The interpretation of these data is as given in Fig. 2. The susceptible plants were presumed to have been the result of inactivation of the Rp1f gene following insertion of Mu1. Since the plants were heterozygous, there was immediate expression of the change in phenotype. It is not yet known whether this interpretation is correct.

A clone of Mu1 is available (22). If the DNA of these plants were digested with a restriction enzyme that does not cut within Mu1, electrophoresed, and then probed with the clone of Mu1, one of the bands that is identified in that it hybridizes to Mu1 should

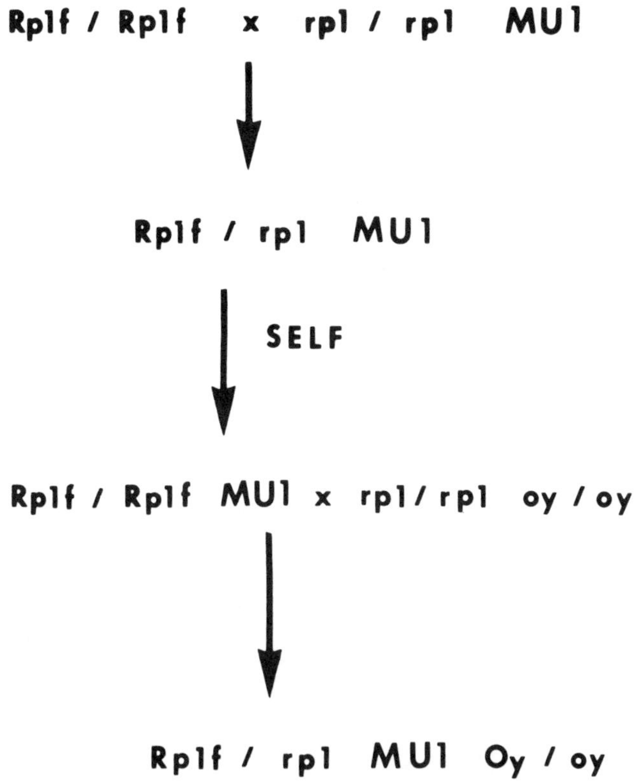

Fig. 1. The crosses used to introduce the transposable element Mu1 into the Rp1f (rust resistance) gene of maize.

also contain flanking sequences that belong to the Rp1f gene. It should be possible to clone the flanking sequences and, if the flanking sequences are unique to Rp1f, use the new clones to probe for the nonmutated, functional Rp1f gene.

The approach of using transposable elements as molecular tags to clone genes for disease resistance is not without its problems. One advantage of Mutator compared to other transposable elements, such as AC-DS, Spm, etc., is that is seems to produce a higher frequency of mutations. It has the disadvantage in that it also is present in several copies. Between 10

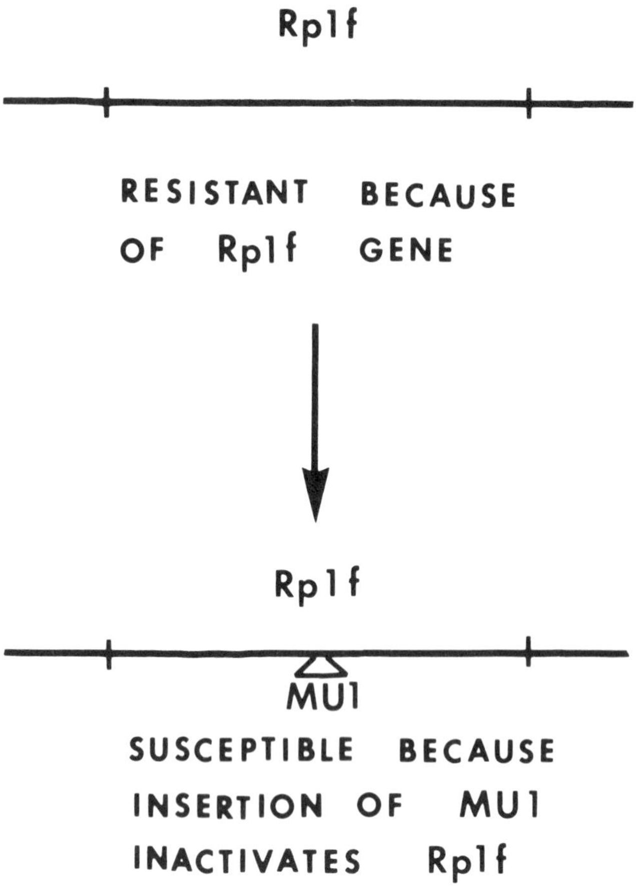
Rp1f
RESISTANT BECAUSE
OF Rp1f GENE
Rp1f
MU1
SUSCEPTIBLE BECAUSE
INSERTION OF MU1
INACTIVATES Rp1f

Fig. 2. The interpretation of the basis for susceptible plants in a population of resistant plants of genotype Rp1f/rp1 Mu1 Oy/oy.

to 60 copies per plant is common. If there are 30 copies per plant, which copy is in Rp1f? This is being investigated, but the high number of copies means more work. With no available genetic transformation system in maize to test directly which insert is in Rp1f, it is necessary to depend on segregations in test crosses to sort the various bands that hybridize to Mu1.

It is not known whether the procedure of

molecular tagging of disease resistance genes with transposons will work. There seems to be no reason why it should not work. The technique has been used with other eukaryotic and prokaryotic organisms. The technique should work for any gene for which a change in phenotype can be scored. It will be interesting to see if the cloning by function will identify the same DNA sequences as cloning by molecular tagging. Cloning by function, however, will have to await the development of a genetic transformation system in maize.

A major objective of much research in plant pathology is to understand the mechanisms of interaction between plant and pathogen. In the past the principal approaches have been through the comparative biochemistry, physiology, ultrastructure, etc., of healthy and diseased, resistant and susceptible, virulent and avirulent, etc., comparisons. From these observations, there was an effort to move back toward the genes that were known to control the differences observed. The opportunity now is to begin to go directly to the gene, determine its product, and move from the gene toward the phenotype. It is an opportunity that seems to hold exceptional promise to sort out which differences observed in the comparative studies are primary and which are secondary, tertiary, etc., effects of the primary event.

For more than 75 years plant pathologists and breeders have manipulated genes for disease resistance for practical gain. But the products of these genes, whose importance to mankind is inestimable, are still unknown. There is now an opportunity, through the use of molecular genetics arguments and recombinant DNA technology, to go directly for the genes that control host-parasite specificity. Genes that control race specificity in several pathogens have already been cloned. The products of these genes should be known soon. The cloning of several disease resistance genes is underway in several laboratories, using several types of arguments and techniques. It seems very likely that this information will have great effects on plant pathology.

What sorts of questions have been asked that will be affected by cloning the genes governing interactions between pathogen and host? A clone of a P or R gene can be sequenced. The probable primary structure of its product can be deduced from the sequence. The product can be made by *in vitro* transcription and translation. The many debates about whether resistance is constitutive or inducible can be resolved. Do the P and R genes determine which plant part is affected by a pathogen? What is the site of gene product in the cell? Is the product in the cell membrane, cytoplasm, organelle, etc.? Is the whole product active in host-pathogen interactions? Are there functions in the cell for the product other than host-parasite interactions? Can clones of the R genes be used to develop genetic transformation systems in plants? Can recombinant DNA techniques be used to transfer R genes from species to species that do not cross by conventional sexual crossing? Can combinations of R genes be assembled that give greater stability to breeding for resistance? The list of questions can go on. The time in the development of recombinant DNA technology is approaching when these questions can be tested directly.

Molecular genetics arguments and recombinant DNA technology are already having an effect on plant pathology. Genes controlling race specificity in bacterial pathogens have already been cloned. Genes controlling disease resistance will be cloned in the near future. Routine transformation systems will be developed for crop plants. Recombinant DNA techniques are already being used for diagnosing virus infections in plants. A large number of biological control agents are already on hand waiting to be tested as soon as permission to test in the field is approved. That plant pathology will not be affected by the recombinant DNA technology (a passing fad?) is already a fallacious statement.

Molecular genetics and the recombinant DNA technology has given a perspective that permits a direct attack on finding the products of the P and R genes. We need to learn to use these arguments and technologies in plant pathology because I believe we would be delinquent in our responsibilities as reflective beings if we do not make use of available resources in our attempt to understand the biological systems with which we work and upon which we are dependent for our livelihood.

LITERATURE CITED

1. Aist, J. R. 1976. Papillae and related wound plugs of plant cells. Annu. Rev. Phytopathol. 14:145-163.
2. Aist, J. R. 1976. Cytology of penetration and infection-fungi. pp. 197-221 in: Physiological Plant Pathology, R. Heitefuss and P. H. Williams, eds. Springer-Verlag, Berlin.
3. Akai, S., Kunoh, H., and Fukutomi, M. 1968. Histochemical changes of barley leaves infected by Erysiphe graminis hordei. Mycopathol. Mycol. Appl. 35:175-180.
4. Akutsu, K., Amano, K., Doi, Y., and Yora, K. 1977. Development of cytoplasmic vesicles around papillae in barley leaf cells following infection with Erysiphe graminis f. sp. hordei. Ann. Phytopathol. Soc. Jpn. 43:491-496.
5. Akutsu, K., Doi, Y., and Yora, K. 1980. Elementary analysis of papillae and cytoplasmic vesicles formed at the penetration site of Erysiphe graminis f. sp. hordei in epidermal cells of barley leaves. Ann. Phytopathol. Soc. Jpn. 46:667-671.
6. Alba, A. P. C., and DeVay, J. E. 1985. Detection of cross-reactive antigens between Phytophthora infestans (Mont.) DeBary and Solanum species by indirect enzyme-linked immunosorbent assay. Phytopath. Z. 112:97-104.
7. Albersheim, P., and Anderson-Prouty, A. J. 1975. Carbohydrates, proteins, cell surfaces, and the biochemistry of pathogenesis. Annu. Rev. Plant Physiol. 26:31-52.
8. Albersheim, P., Darvill, A. G., Davis, K. R., Lau, J. M., McNeil, M., Sharp, J. K., and York, W. S. 1984. Why study the structures of

biological molecules? The importance of studying the structures of complex carbohydrates. pp. 19-51 in: Structure, Function and Biosynthesis of Plant Cell Walls. W. M. Dugger and S. Bartnicki-Garcia, eds. American Soc. Plant Physiologists, Rockville, Maryland.

9. Anderson, A. J. 1980. Differences in the biochemical compositions and elicitor activity of extracellular components produced by three races of a fungal plant pathogen, Colletotrichum lindemuthianum. Can. J. Microbiol. 26:1473-1479.

10. Anikster, Y., and Wahl, I. 1979. Coevolution of the rust fungi on Gramineae and Liliaceae and their hosts. Annu. Rev. Phytopathol. 17:367-403.

11. Asada, Y., Ohguchi, T., and Matsumoto, I. 1979. Induction of lignification in response to fungal infection. pp. 99-112 in: Recognition and Specificity in Plant Host-parasite Interactions. J. M. Daly and I. Uritani, eds. Japan Scientific Societies Press, Tokyo, and University Park Press, Baltimore.

12. Ayers, A. R., Ebel, J., Valent, B., and Albersheim, P. 1976. Host-pathogen interactions. X. Fractionation and biological activity of an elicitor isolated from the mycelial walls of Phytophthora megasperma var. sojae. Plant Physiol. 57:760-765.

13. Bailey, J. A. 1980. Constitutive elicitors from Phaseolus vulgaris, a possible cause of phytoalexin accumulation. Ann. Phytopathol. 12:395-402.

14. Bailey, J. A. 1982. Physiological and biochemical events associated with the expression of resistance to disease. pp 39-65 in: Active Defense Mechanisms in Plants. R. K. S. Wood, ed. Plenum Press, New York.

15. Bailey, J. A., and J. W. Mansfield. 1982. Phytoalexins. John Wiley, New York.

16. Baker, H. G. 1972. Human influences on plant evolution. Economic Botany 26:32-43.

17. Ball, E. 1969. Histology of mixed callus cultures. Bull. Torrey Bot. Club 96:52-59.

18. Ball, E. A. 1971. Growth of plant tissue upon a

substrate of another kind of tissue. 1. Qualitative observations. Z. Pflanzenphysiol. 65:140-158.

19. Beardmore, J., Ride, J. P., and Granger, J. W. 1983. Cellular lignification as a factor in the hypersensitive resistance of wheat to stem rust. Physiol. Plant Pathol. 22:209-220.

20. Bell, J. N., Dixon, R. A., Bailey, J. A., Rowell, P. M., and Lamb, C. J. 1984. Differential induction of chalcone synthase mRNA activity at the onset of phytoalexin accumulation in compatible and incompatible plant-pathogen interactions. Proc. Nat. Acad. Sci. (U.S.A.) 81:3384-3388.

21. Benada, J. 1970. Observations on early phases of infection by powdery mildew (Erysiphe graminis D.C.). Phytopathol. Z. 68:181-187.

22. Bennetzen, J. L., Swanson, J., Taylor, W. C., and Freeling, M. 1984. DNA insertion in the first intron of maize Adhl affects message levels: cloning of progenitor and mutant Adhl alleles. Proc. Natl. Acad. Sci. (U.S.A.) 81:4125-4128.

23. Biggs, D. R., Lane, G. A., Russell, G. B., and Sutherland, O. R. W. 1983. Insect feeding deterrent activity of isoflavonoid phytoalexins. Abstr. Fourth Int. Congr. Plant Pathol. pp. 176.

24. Bloom, B. R. 1979. Games parasites play: How parasites evade immune surveillance. Nature 279:21-26.

25. Briggle, L. W. 1966. Three loci in wheat involving resistance to Erysiphe graminis f. sp. tritici. Crop Science 6:461-465.

26. Brown, K. N. 1976. Specificity in host-parasite interaction. pp. 120-175 in: Receptors and Recognition. Vol. 1, Series A. P. Cuartrescasas and M. F. Greaves, eds. Halstead Press, John Wiley and Sons, New York.

27. Bruegger, B. B., and Keen, N. T. 1979. Specific elicitors of glyceollin accumulation in the Pseudomonas glycinea-soybean host-parasite system. Physiol. Plant Pathol. 15:43-51.

28. Bushnell, W. R. 1971. The haustorium of Erysiphe graminis: an experimental study by

light microscopy. pp. 229-254 in: Morphological and Biochemical Events in Plant-Parasite Interaction. S. Akai and S. Ouchi, eds. Phytopathological Society of Japan, Tokyo.

29. Bushnell, W. R. 1972. Physiology of fungal haustoria. Annu. Rev. Phytopathol. 10:151-176.

30. Bushnell, W. R. 1976. Reactions of cytoplasm and organelles in relation to host-parasite specificity. pp. 131-150 in: Specificity in Plant Disease. R. K. S. Wood and A. Graniti, eds. Plenum Press, New York.

31. Bushnell, W. R. 1979. The nature of basic compatibility: comparisons between pistil-pollen and host-parasite interaction. pp. 221-227 in: Recognition and Specificity in Plant Host-Parasite Interactions. J. M. Daly and I. Uritani, eds. Japan Scientific Societies Press, Tokyo, and University Park Press, Baltimore.

32. Bushnell, W. R., and Rowell, J. B. 1981. Supressors of defense reactions: a model for roles in specificity. Phytopathology 71:1012-1014.

33. Bushnell, W. R., and Zeyen, R. J. 1976. Light and electron microscope studies of cytoplasmic aggregates formed in barley cells in response to Erysiphe graminis. Can. J. Bot. 54:1647-1655.

34. Bushnell, W. R., Dueck, J., and Rowell, J. B. 1967. Living haustoria and hyphae of Erysiphe graminis f. sp. hordei with intact and partly dissected host cells of Hordeum vulgare. Can. J. Bot. 45:1719-1732.

35. Callow, J. A. 1977. Recognition, resistance and the role of plant lectins in host-parasite interactions. Adv. Bot. Res. 4:1-49.

36. Callow, J. A. 1984. Cellular and molecular recognition between higher plants and fungal pathogens. pp. 212-237 in: Cellular Interactions. Encyclopedia of Plant Physiology, New Series Vol. 17. H. F. Linskens and J. Heslop-Harrison, eds. Springer-Verlag, Berlin.

37. Carlile, M. J. 1972. The lethal interaction following plasmodial fusion between two strains of the myxomycete Physarum polycephalum. J. Gen.

Microbiol. 71:581-590.
38. Carver, T. L. W. 1983. Avoidance of host resistance to primary penetration by barley powdery mildew through establishment of endophytic mycelia. Rep. Welsh Plant Breed. Stn. for 1982, p. 182.
39. Carver, T. L. W. 1985. Host recognition by *Erysiphe graminis* conidia during the early stages of germination. Rep. Welsh Plant Breed. Stn. for 1984, pp. 158-162.
40. Carver, T. L. W., and Bushnell, W. R. 1983. The probable role of primary germ tubes in water uptake before infection by *Erysiphe graminis.* Physiol. Plant Pathol. 23:229-240.
41. Carver, T. L. W., and Carr, A. J. H. 1977. Race non-specific resistance of oats to primary infection by mildew. Ann. Appl. Biol. 86:29-36.
42. Carver, T. L. W., and Carr, A. J. H. 1978. Effects of host resistance on the development of haustoria and colonies of oat mildew. Ann. Appl. Biol. 88:171-178.
43. Carver, T. L. W., and Carr, A. J. H. 1978. The early stages of mildew colony development on susceptible oats. Ann. Appl. Biol. 89:201-209.
44. Carver, T. L. W., and Chamberlain, J. A. 1978. Scanning electron microscopy of barley and oat mildew haustoria. Trans. Br. Mycol. Soc. 70:30-31.
45. Carver, T. L. W., and Phillips, M. 1982. Effects of photoperiod and level of irradiance on production of haustoria by *Erysiphe graminis* f. sp. *hordei*. Trans. Br. Mycol. Soc. 79:207-211.
46. Carver, T. L. W., and Williams, O. 1980. The influence of photoperiold on growth patterns of *Erysiphe graminis* f. sp. *hordei.* Ann. Appl. Biol. 94:405-414.
47. Carver, T. L. W., Zeyen, R. J., and Ahlstrand, G. G. 1982. Sequential use of light microscopy, scanning electron microscopy and X-ray microanalysis to study powdery mildew infection. Phytopathology 72:968 (Abstr.).
48. Charudattan, R., and DeVay, J. E. 1981. Purification and partial characterization of an

antigen from Fusarium oxysporum f. sp. vasinfectum that cross-reacts with antiserum to cotton (Gossypium hirsutum) root antigens. Physiol. Plant Pathol. 18:289-295.

49. Chilton, M. D., Drummond, M. H., Merlo, D. J., Sciaky, D., Montoya, A. L., Gordon, M. P., and Nester, E. W. 1977. Stable incorporation of plasmid DNA into higher plant cells: The molecular basis of crown gall tumorigenesis. Cell 11:263-271.

50. Chong, J., and Harder, D. E. 1982. Ultrastructure of haustorium development in Puccinia coronata f. sp. avenae: cytochemistry and energy dispersive X-ray analysis of the haustorial mother cells. Phytopathology 72:1518-1526.

51. Clifford, B. C. 1970. Brown rust of barley. Welsh Plant Breed. Stn. Jubilee Rep. for 1919-1969, pp. 124-126.

52. Clifford, B. C. 1972. The histology of race non-specific resistance to Puccinia hordei Otth. in barley. in: Proc. European and Medn. Cereal Rusts Conf., Prague 1972 1:75-79.

53. Clifford, B. C. 1974. Relationship between compatible and incompatible infection sites of Puccinia hordei on barley. Trans. Br. Mycol. Soc. 63:215-220.

54. Clifford, B. C. 1985. Barley leaf (brown) rust. in: The Cereal Rusts, Vol. 2. A. P. Roelfs and W. R. Bushnell, eds. Academic Press, New York (in press).

55. Clifford, B. C., and Clothier, R. B. 1974. Physiologic specialization of Puccinia hordei on barley hosts with non-hypersensitive resistances. Trans. Br. Mycol. Soc. 63:421-430.

56. Clifford, B. C., and Roderick, H. W. 1981. Detection of cryptic resistance of barley to Puccinia hordei. Trans. Br. Mycol. Soc. 76:17-24.

57. Clifford, B. C., and Schafer, J. F. 1966. Some effects of temperature on the expression of mature tissue resistance of oats to crown rust. Phytopathology 56:874 (Abstr.).

58. Collins, O. R., and Betterly, D. A. 1982.

Didymium iridis in past and future research. pp. 25-57 in: Cell Biology of Physarum and Didymium, Vol. 1. H. C. Aldrich, and J. W. Daniel, eds. Academic Press, London, New York.
59. Cooke, R. C., and Whipps, J. M. 1980. The evolution of modes of nutrition in fungi parasitic on terrestrial plants. Biol. Rev. 55:341-362.
60. Daly, J. M. 1972. The use of near-isogenic lines in biochemical studies of the resistance of wheat to stem rust. Phytopathology 62:392-400.
61. Daly, J. M. 1976. Specific interactions involving hormonal and other changes. pp. 151-167 in: Specificity in Plant Diseases. R. K. S. Wood and A. Graniti, eds. Plenum Press, New York and London.
62. Daly, J. M., and Knoche, H. W. 1982. The chemistry and biology of pathotoxins exhibiting host-selectivity. Adv. Plant Pathol. 1:83-138.
63. Damian, R. T. 1979. Molecular mimicry in biological adaptation. pp. 103-126 in: Host Parasite Interfaces. B. B. Nickol, ed. Academic Press, New York.
64. Darvill, A. G., and Albersheim, P. 1984. Phytoalexins and their elicitors. Annu. Rev. Plant Physiol. 35:243-275.
65. Dausset, J. 1981. The major histocompatibility complex in man. Past, present and future concepts. Science 213:1469-1474.
66. Davey, M. R., and Kumar, A. 1983. Higher plant protoplasts - retrospect and prospects. pp. 219-299 in: Plant Protoplasts, Int. Rev. Cytol. Supp. 16. K. L. Giles, ed. Academic Press, New York and London.
67. Davey, M. R., and Power, J. B. 1975. Polyethylene glycol-induced uptake of micro-organisms into higher plant protoplasts: an ultrastructural study. Plant Sci. Lett. 5:269-274.
68. Davey, M. R., Clothier, R. H., Balls, M., and Cocking, E. C. 1978. An ultrastructural study of the fusion of cultured amphibian cells with higher plant protoplasts. Protoplasma 96:157-172.
69. Davis, K. R., Lyon, G. D., Darvill, A. G., and

Albersheim, P. 1984. Host-pathogen interactions. XXV. Endopolygalacturonic acid lyase from *Erwinia carotovora* elicits phytoalexin accumulation by releasing plant cell wall fragments. Plant Physiol. 74:52-60.

70. Day, P. R. 1974. Genetics of Host-Parasite Interaction. W. H. Freeman and Company, San Francisco.

71. Depicker, A., Van Montagu, M., and Schell, J. 1982. Plant cell transformation by *Agrobacterium* plasmids. in: Genetic Engineering of Plants. T. Kosuge, C. Meredith and A. Hollander eds. Plenum Press, New York and London.

72. DeVay, J. E., and Adler, H. E. 1976. Antigens common to hosts and parasites. Annu. Rev. Microbiol. 30:147-168.

73. DeVay, J. E., Schnathorst, W. C., and Foda, M. S. 1967. Common antigens and host-parasite interactions. pp. 313-328 in: The Dynamic Role of Molecular constituents in Plant-Parasite Interactions. C. J. Mirocha and I. Uritani, eds. American Phytopathological Society, St. Paul.

74. DeVay, J. E., Wakeman, R. J., Kavanagh, J. A., and Charudattan, R. 1981. The tissue and cellular location of major cross-reactive antigen shared by cotton and soil-borne fungal parasites. Physiol. Plant Pathol. 18:59-66.

75. DeWit, P. J. G. M., and Spikman, G. 1982. Evidence for the occurrence of race and cultivar-specific elicitors of necrosis in intercellular fluids of compatible interactions of *Cladosporium fulvum* and tomato. Physiol. Plant Pathol. 21:1-11.

76. DeWit, P. J. G. M., Hofman, A. E., Velthuis, G. C. M., and Kuc, J. A. 1984. Isolation and characterization of an elicitor of necrosis isolated from intercellular fluids of compatible interactions of *Cladosporium fulvum* (syn. *Fulvia fulva*) and tomato. Plant Physiol. 77:642-647.

77. Dixon, R. A., Dey, P. M., Lawton, M. A., and Lamb C. J. 1983. Phytoalexin induction in French bean. Intercellular transmission of elicitation in cell suspension cultures and hypocotyl

sections of Phaseolus vulgaris. Plant Physiol. 71:251-256.

78. Doke, N., and Tomiyama, K. 1980. Suppression of the hypersensitive response of potato tuber protoplasts to hyphal wall components by water soluble glucans isolated from Phytophthora infestans. Physiol. Plant Pathol. 16:177-186.

79. Dudits, D., Rasko, I., Hadlaczky, G., and Lima-de-Faria, A. 1976. Fusion of human cells with carrot protoplasts induced by polyethylene glycol. Hereditas 82:121-124.

80. Ebel, J., Schmidt, W. E., and Loyal, R. 1984. Phytoalexin synthesis in soybean cells: elicitor induction of phenylalanine ammonia-lyase and chalcone synthase mRNAs and correlation with phytoalexin accumulation. Arch. Biochem. Biophys. 232:240-248.

81. Ellingboe, A. H. 1972. Genetics and physiology of primary infection by Erysiphe graminis. Phytopathology 62:401-406.

82. Ellingboe, A. H. 1976. Genetics of host-parasite interactions. pp. 761-778 in: Encyclopedia of Plant Physiology, New Series, Vol. 4:Physiological Plant Pathology. R. Heitefuss and P. H. Williams, eds. Springer-Verlag, Berlin, Heidelberg and New York.

83. Ellingboe, A. H. 1978. A genetic analysis of host-parasite interactions. pp. 159-180 in: The Powdery Mildews. D. M. Spencer, ed. Academic Press, London.

84. Ellingboe, A. H. 1982. Genetical aspects of active defense. pp. 179-192 in: Active Defense Mechanisms in Plants. R. K. S. Wood, ed. Plenum Press, New York and London.

85. Ellingboe, A. H., and Gabriel, D. W. 1977. Induced conditional mutants for studying host/pathogen interactions. in: Induced Mutations Against Plant diseases, IAEA, Vienna.

86. Elmhirst, J. F., and Heath, M. C. 1984. Host-parasite interactions of Uromyces phaseoli var. typica and U. phaseoli var. vigne with species of the Phaseolus - Vigna plant complex. Phytopathology 74:850 (Abstr.)

87. Esser, K., and Meinhardt, F. 1984. Barrage formation in fungi. pp 350-361 in: Cellular Interactions. Encyclopedia of Plant Physiology, New Series Vol. 17. H. F. Linskens, and J. Heslop-Harrison, eds. Springer-Verlag, Berlin.
88. Ferenczy, L. 1980. Fusion of protoplasts of auxotrophic fungal mutants: diversity in the genetic background of nutritional complementation. pp. 55-62 in: Advances in Protoplast Research. L. Ferenczy and G. L. Farkas eds. Pergamon, Oxford.
89. Flor, H. H. 1946. Genetics of pathogenicity in Melampsora lini. J. Agric. Res. 73:335-337.
90. Flor, H. H. 1947. Inheritance of reaction to rust in flax. J. Agric. Res. 74:241-262.
91. Flor, H. H. 1971. Current status of the gene-for-gene concept. Annu. Rev. Phytopathol. 9:275-296.
92. Fowke, L. C., Gresshoff, P. M., and Marchant, H. J. 1979. Transfer of organelles of the alga Chlamydomonas reinhardii into carrot cells by protoplast fusion. Planta 144:341-347.
93. Fowke, L. C., Marchant, H. J., and Gresshoff, P. M. 1981. Fusion of protoplasts from carrot cell cultures and the green alga Stigeocloium. Can. J. Bot. 59:1021-1025.
94. Fujii, T., and Nito, N. 1972. Studies on the compatibility and grafting of fruit trees. I. Callus fusion between rootstock and scion. J. Japan Soc. Hort. Sci. 41:1-10.
95. Gabriel, D. W. 1985. Molecular cloning of specific avirulence genes from Xanthomonas malvacearum. Proc. 2nd Intern. Symp. Molecular Genetics of Bacteria-Plant Interactions. (In press).
96. Gautheret, R. J. 1945. Une Voie Nouvele en Biologie Végetale. La Culture des Tissues. L'Avenir de la Science-21. Gallimard.
97. Germar, B. 1935. On some effects of silicic acid on cereal plants especially on their resistance to mildew. Rev. Appl. Mycol. 14:25 (Abstr.).
98. Gerson, D. F., Meadows, M. G., Finkelman, M., and

Walden, D. B. 1980. The biophysics of protoplast fusion. pp. 447-462 in: Advances in Protoplast Research. L. Ferenczy and G. L. Farkas, eds. Pregamon, Oxford.

99. Goodman, R. N. 1976. Physiological and cytological aspects of the bacterial infection process. pp. 172-196 in: Encyclopedia of Plant Physiology, New Series, Vol 4:Physiological Plant Pathology. R. Heitefuss and P. H. Williams, eds. Springer-Verlag, Berlin, Heidelberg and New York.

100. Graf Marin, A. 1934. Studies on powdery mildew of cereal. Mem. Cornell. Univ. Agric Exp. Stn. 157:1-48.

101. Graham, B. F. Jr., and Bormann, F. H. 1966. Natural root grafts. Bot. Rev. 32:255-292.

102. Hadlaczky, G., Burg, K., Maroy, P., and Dudits, D. 1980. DNA synthesis and division in interkingdom heterokaryons. In Vitro 16:647-650.

103. Hagmann, M. L., Heller, W., and Grisebach, H. 1984. Induction of phytoalexin synthesis in soybean. Stereospecific 3,9-dihydroxypterocarpan 6a-hydroxylase from elicitor-induced soybean cell cultures. Eur. J. Biochem. 142:127-131.

104. Hahn, M. G., Bonhoff, A., and Grisebach, H. 1985. Quantitative localization of the phytoalexin glyceollin I in relation to fungal hyphae in soybean roots infected with _Phytophthora megasperma_ f. sp. _glycinea_. Plant Physiol. 77:591-601.

105. Harder, D. E., and Chong, J. 1984. Structure and physiology of haustoria. pp. 431-476 in: The Cereal Rusts, Vol. 1. W. R. Bushnell and A. P. Roelfs, eds. Academic Press, New York.

106. Hartmann, H. T., and Kester, D. E. 1968. Plant Propagation: Principles and Practices. 2nd edition. Prentice-Hall, Englewood Cliffs, N. J.

107. Hasezawa, S., Nagata, T., and Syono, K. 1981. Transformation of _Vinca_ protoplasts mediated by _Agrobacterium_ spheroplasts. Mol. Gen. Genet. 182:206-210.

108. Heath, M. C. 1974. Light and electron microscope studies of the interactions of host and non-host plants with cowpea rust-_Uromyces_

phaseoli var. vignae. Physiol. Plant Pathol. 4:403-414.
109. Heath, M. C. 1977. A comparative study of non-host interactions with rust fungi. Physiol. Plant Pathol. 10:73-88.
110. Heath, M. C. 1979. Partial characterization of the electron-opaque deposits formed in the non-host plant, French bean, after cowpea rust infection. Physiol. Plant. Pathol. 15:141-148.
111. Heath, M. C. 1979. Effects of heat shock, actinomycin D, cycloheximide and blasticidin S on nonhost interactions with rust fungi. Physiol. Plant Pathol. 15:211-218.
112. Heath, M. C. 1980. Reactions of nonsuscepts to fungal pathogens. Annu. Rev. Plant Pathol. 18:211-236.
113. Heath, M. C. 1981. The suppression of the development of silicon containing deposits in french bean leaves by exudates of the bean rust fungus and extracts from bean rust-infected tissue. Physiol. Plant Pathol. 18:149-155.
114. Heath, M. C. 1981. Nonhost resistance. pp. 201-217 in: Plant Disease Control: Resistance and Susceptibility. R. C. Staples and G. H. Toenniessen, eds. John Wiley and Sons Inc. New York.
115. Heath, M. C. 1981. Insoluble silicon in necrotic cowpea cells following infection with an incompatible isolate of the cowpea rust fungus. Physiol. Plant Pathol. 19:273-276.
116. Heath, M. C. 1981. A generalized concept of host-parasite specificity. Phytopathology 71:1121-1123.
117. Heath, M. C. 1982. The absence of active defense mechanisms in compatible host-pathogen interactions. NATO Adv. Study Inst. Ser., Ser. A. 37:143-156.
118. Heslop-Harrison, J. 1975. Incompatibility and the pollen-stigma interaction. Annu. Rev. Plant Physiol. 26:403-425.
119. Hijwegen, T. 1979. Fungi as plant taxonomists. Symb. Bot. Ups. XXII 4:146-165.
120. Hirata, K. 1959. Preliminary report concerning

the effects of lithium on barley powdery mildew (Erysiphe graminis f. sp. hordei) and the host tissue. Bull. Fac. Agr., Niigata Univ. 11:34-42.

121. Hirata, K. 1960. Observations on the development of young colonies of the barley powdery mildew (Erysiphe graminis hordei Marchal.) Trans. Mycol. Soc. Jpn. 3:43-66.

122. Hirata, K. 1967. Notes on haustoria, hyphae and conidia of the powdery mildew fungus of barley, Erysiphe graminis f. sp. hordei. Mem. Fac. Agric., Niigata Univ. 6:207-259.

123. Hirata, K. 1971. Calcium in relation to the susceptiblity of primary barley leaves to powdery mildew. pp. 207-228 in: Morphological and Biochemical Events in Plant-Parasite Interaction, S. Akai and S. Ouchi, eds. Phytopathological Society of Japan, Tokyo.

124. Hogenboom, N. G. 1975. Incompatibility and incongruity: two different mechanisms for the non-functioning of intimate partner relationships. Proc. Roy. Soc. (London) B. 188:361-375.

125. Hogenboom, N. G. 1983. Bridging a gap between related fields of research: pistil-pollen relationships and the distinction between incompatibility and incongruity in nonfunctioning host-parasite relationships. Phytopathology 73:381-385.

126. Hogenboom, N. G. 1984. Incongruity: nonfunctioning of intercellular and intracellular partner relationships through non-matching information. pp. 640-654 in: Cellular Interactions, Encyclopedia of Plant Physiology, New Series Vol. 17. H. F. Linskens and J. Heslop-Harrison, eds. Springer-Verlag, Berlin.

127. Homma, Y. 1937. Erysiphaceae of Japan. Rev. Appl. Mycol. 16:633 (Abstr.).

128. Ingram, D. S. 1982. A structural view of active defense. pp. 19-38 in: Active Defense Mechanisms in Plants. R. K. S. Wood, ed. Plenum, New York and London.

129. Johnson, L. E. B., Bushnell, W. R., and Zeyen, R. J. 1979. Binary pathways for analysis of primary infection and host response in popula-

tions of powdery mildew fungi. Can. J. bot. 57:497-511.
130. Johnson, L. E. B., Bushnell, W. R., and Zeyen, R. J. 1982. Defense patterns in nonhost higher plant species against two powdery mildew fungi. I. Monocotyledonous species. Can. J. Bot. 60:1068-1083.
131. Jones, C. W., Mastrangelo, I. A., Smith, H. H., Liu, H. Z., and Meck, R. A. 1976. Interkingdom fusion between human (HeLa) cells and tobacco hybrid (GGLL) protoplasts. Science 193:401-403.
132. Jones, I. T., and Hayes, J. D. 1971. The effect of sowing date on adult plant resistance to Erysiphe gramnis f. sp. avenae in oats. Ann. Appl. Biol. 68:31-39.
133. Jones, I. T., O'Reilly, A., and Davies, I. J. E. R. 1983. Inheritance of adult plant resistance to oat mildew. Rep. Welsh Plant Breed. Stn. for 1982, pp. 103-105.
134. Jorgensen, J. H., and Mortensen, K. 1977. Primary infection by Erysiphe graminis f. sp. hordei of barley mutants with resistance genes in the Ml-o locus. Phytopathology 67:678-685.
135. Katz, D. H. 1978. Self recognition as a means of communication in the immune system. pp. 411-428 in: Cell-Cell Recognition, Symp. Soc. Exp. Biol. Vol. 33. A. S. G. Curtis, ed. Cambridge Univ. Cambridge.
136. Keen, N. T. 1981. Evaluation of the role of phytoalexins. pp. 155-177 in: Plant Disease Control. R. Staples and G. Toenessien, eds. John Wiley, New York.
137. Keen, N. T. 1982. Specific recognition in gene-for-gene host-parasite systems. Adv. Plant Pathol. 1:35-82.
138. Keen, N. T., and Holliday, M. J. 1982. Recognition of bacterial pathogens by plants. pp. 179-217 in: Phytopathogenic Prokaryotes, Vol. 2. M. Mount and G. Lacy, eds. Academic Press, New York.
139. Keen, N. T., and Legrand, M. 1980. Surface glycoproteins: evidence that they may function as the race specific phytoalexin elicitors of

Phytophthora magasperma f. sp. glycinea. Physiol. Plant Pathol. 17:175-192.

140. Keen, N. T., and Yoshikawa, M. 1983. β-1,3-endoglucanase from soybean releases elicitor-active carbohydrates from fungus cell walls. Plant Physiol. 71:460-465.

141. Keen, N. T., Ersek, T., Long, M., Bruegger, B., and Holliday, M. 1981. Inhibition of the hypersensitive reaction of soybean leaves to incompatible Pseudomonas spp. by blasticidin S, streptomycin or elevated temperature. Physiol. Plant Pathol. 18:325-337.

142. Keen, N. T., Midland, S., and Sims, J. J. 1983. Purification of victorin. Phytopathology 73:830. (Abstr.).

143. Kidger, A. L., and Carver, T. L. W. 1981. Autofluorescence in oats infected by powdery mildew. Trans. Br. Mycol. Soc. 76:405-409.

144. Kúc, J., Tjamos, E., and Bostock, R. 1984. Metabolic regulation of terpenoid accumulation and disease resistance in potato. pp. 103-126 in: Isopentenoids in Plants, Biochemistry and Function. W. D. Nes, G. Fuller, and L-S. Tsai, eds. Mercel Dekker, New York.

145. Kunoh, H. 1981. Early stages of infection process of Erysiphe graminis on barley and wheat. pp. 85-101 in: Microbial Ecology of the Phylloplane. J. P. Blakeman, ed. Academic Press, London.

146. Kunoh, H., and Ishizaki, H. 1975. Silicon levels near penetration sites of fungi on wheat, barley, cucumber and morning glory leaves. Physiol. Plant Pathol. 5:283-287.

147. Kunoh, H., and Ishizaki, H. 1976. Accumulation of chemical elements around penetration sites of Erysiphe graminis hordei on barley leaf epidermis (II). Level of silicon in papilla around the haustorial neck. Ann. Phytopathol. Soc. Jpn. 42:30-34.

148. Kunoh, H., and Ishizaki, H. 1976. Accumulation of chemical elements around the penetration sites of Erysiphe graminis hordei on barley leaf epidermis (III). Micromanipulation and X-ray

microanalysis of silicon. Physiol. Plant Pathol. 8:91-96.

149. Kunoh, H., and Ishizaki, H. 1981. Cytological studies of early stages of powdery mildew in barley and wheat. VII. Reciprocal translocation of a fluorescent dye between barley coleoptile cells and conidia. Physiol. Plant Pathol. 18:207-211.

150. Kunoh, H., Ishizaki, H., and Kondo, F. 1975. Composition analysis of 'halo' area of barley leaf epidermis induced by powdery mildew infection. Ann. Phytopathol. Soc. Jpn. 41:33-39.

151. Kunoh, H., Takamatsu, S., and Ishizaki, H. 1978. Cytological studies of early stages of powdery mildew in barley and wheat. III. Distribution of residual calcium and silicon in germinated conidia of _Erysiphe graminis hordei._ Physiol. Plant Pathol. 13:319-325.

152. Kunoh, H., Tsuzuki, T., and Ishizaki, H. 1978. Cytological studies of early stages of powdery mildew in barley and wheat. IV. Direct ingress from superficial primary germ tubes and appressoria of _Erysiphe graminis hordei_ on barley leaves. Physiol. Plant Pathol. 13:327-333.

153. Kunoh, H., Aist, J. R., and Israel, H. W. 1979. Primary germ tubes and host cell penetrations from appressoria of _Erysiphe graminis hordei._ Ann. Phytopathol. Soc. Jpn. 45:326-332.

154. Kunoh, H., Itoh, O., Kohno, M., and Ishizaki, H. 1979. Are primary germ tubes of conidia unique to _Erysiphe graminis?_ Ann. Phytopathol. Soc. Jpn. 45:675-682.

155. Kunoh, H., Yamamori, K., and Ishizaki, H. 1982. Cytological studies of early stages of powdery mildew in barley and wheat. VIII. Autofluorescence at penetration sites of _Erysiphe graminis hordei_ on living barley coleoptiles. Physiol. Plant Pathol. 21:373-379.

156. Lacy, M. L., and Horner, C. E. 1966. Behaviour of _Verticillium dahliae_ in the rhizosphere and on roots of plants susceptible, resistant, and immune to wilt. Phytopathology 56:427-430.

157. Lane, E. B., and Carlile, M. J. 1979. Post-

fusion somatic incompatibility in plasmodia of *Physarum polycephalum*. J. Cell Sci. 35:339-354.
158. Lee, S. C., and West, C. A. 1981. Polygalacturonase from *Rhizopus stolonifer*, an elicitor of casbene synthetase activity in castor bean (*Ricinus communis* L.) seedlings. Plant Physiol. 67:633-639.
159. Leube, J., and Grisebach, H. 1983. Further studies on induction of enzymes of phytoalexin synthesis in soybean and cultured soybean cells. Z. Naturforsh 38c:730-735.
160. Lima-de-Faria, A., Eriksson, T., and Kjellen, L. 1977. Fusion of human cells with *Haplopappus* protoplasts by means of Sendai virus. Hereditas 87:57-66.
161. Lin, M.-R., and Edwards, H. H. 1974. Primary penetration process in powdery mildewed barley related to host cell age, cell type and occurrence of basic staining material. New Phytol. 73:131-137.
162. Littlefield, L. J., and Heath, M. C. 1979. Ultrastructure of the rust fungi. Academic Press, New York.
163. Litzenberger, S. C. 1949. Nature of susceptibility to *Helminthosporium victoriae* and resistance to *Puccinia coronata* in Victoria oats. Phytopathology 39:300-318.
164. Long, M., Barton-Willis, P., Staskawicz, B. J., Dahlbeck, D., and Keen, N. T. 1985. Further studies on the relationship between glyceollin accumulation and the resistance of soybean leaves to *Pseudomonas syringae* pv. *glycinea*. Phytopathology 75:235-239.
165. Long, S. R., Meade, H. M., Brown, S. E., and Ausubel, F. M. 1982. Transposon-induced symbiotic mutants of *Rhizobium meliloti*. pp. 129-143 in: Genetic Engineering in the Plant Sciences. N. Panopoulos ed. Praeger Press, New York.
166. Lupton, F. G. H. 1956. Resistance mechanisms of *Triticum*, *Aegilops* and of amphidiploids between them to *Erysiphe graminis* D.C. Trans. Br. Mycol. Soc. 39:51-59.

167. Maniara, G., Laine, R., and Kúc, J. 1984. Oligosaccharides from Phytophthora infestans enhance the elicitation of sesquiterpenoid stress metabolites by arachidonic acid in potato. Physiol. Plant Pathol. 24:177-186.
168. Manners, J. M., and Gay, J. L. 1983. The host parasite interface and nutrient transfer in biotrophic parasitism. pp. 164-195 in: Biochemical Plant Pathology. J. A. Callow, ed. Wiley, New York.
169. Mastrangelo, I. A., and Mitra, J. 1981. Chinese hamster ovary chromosomes and antigens in tobacco/ hamster heterokaryons. J. Hered. 72:81-86.
170. Matta, A. 1982. Mechanisms in non-host resistance. pp. 119-141 in: Active Defense Mechanisms in Plants. R. K. S. Wood, ed. Plenum, New York and London.
171. Mayama, S. 1983. The role of avenalumin in the resistance of oats to crown rust. Mem. Fac. Agric. Kagawa Univ. No. 42.
172. Mayama, S., and Keen, N. T. 1984. Production of victorin by various isolates of Helminthosporium victoriae and its phytoalexin elicitor activity. Phytopathology 14:850. (Abstr.).
173. Mayama, S., and Tani, T. 1981. Relationship between the production of phytoalexin avenalumin and host selective infection of oat Pc line by Helminthosporium victoriae. Ann. Phytopathol. Soc. Japan 47:124. (Abstr.) (In Japanese).
174. Mayama, S., Daly, J. M., and Rehfield, D. W. 1975. Hypersensitive response of near-isogenic wheat carrying the temperature-sensitive Sr6 allele for resistance to stem rust. Physiol. Plant Pathol. 7:35-47.
175. McIntosh, R. A. 1976. Genetics of wheat and wheat rusts since Farrer. J. Austr. Inst. Agric. Sci. 42:203-216.
176. McKeen, W. E., and Rimmer, S. R. 1973. Initial penetration process in powdery mildew infection of susceptible barley leaves. Phytopathology 63:1047-1053.
177. Moore, R., and Walker, D. B. 1981a. Studies of

vegetative compatibility-incompatilbity in higher plants. I. A structural study of a compatible autograft in Sedum telephoides (Crassulaceae). Amer. J. Bot. 68:820-830.

178. Moore, R., and Walker, D. B. 1981b. Studies of vegetative compatibility-incompability in higher plants. II. A structural study of an incompatible heterograft between Sedum telephoides (Crassulaceae) and Solanum pennellii (Solanaceae). Amer. J. Bot. 68:831-842.

179. Murphy, H. C. 1935. Physiologic specialization in Puccinia coronata avenae. U.S. Dept. Agr. Tech. Bull. No. 433.

180. Newton, M., Johnson, T., and Brown, A. M. 1933. Stripe rust in Canada. Phytopathology 23:25-26.

181. Niks, R. E. 1982. Early abortion of colonies of leaf rust, Puccinia hordei in partially-resistant barley seedlings. Can. J. Bot. 60:714-723.

182. Niks, R. E. 1983. Haustorium formation by Puccinia hordei in leaves of hypersensitive, partially resistant and non-host plant genotypes. Phytopathology 73:64-66.

183. Nothnagel, E. A., McNeil, M., Albersheim, P., and Dell, A. 1983. Host-pathogen interactions. XXII. A galacturonic acid oligosaccharide from plant cell walls elicits phytoalexins. Plant Physiol. 71:916-926.

184. Nover, I., and Lehmann, C. O. 1974. Resistenzeigenschaften in Gersten-und Weizensortiment Gatersleben 18. Prüfung von Sommergersten auf ihr Verhalten gegen Zwergrost (Puccinia hordei Otth). Kulturpflanze 22:25-43.

185. Oku, H., Shiraishi, T., and Ouchi, S. 1977. Suppression of induction of phytoalexin, pisatin, by low-molecular weight substances from spore germination fluid of pea pathogen, Mycosphaerella pinodes. Naturwissenschaften 64:643-644.

186. Oku, H., Shiraishi, T., Ouchi, S., Ishiura, M., and Matsueda, R. 1980. A new determinant of pathogenicity in plant disease. Naturwissenschaften 67:310.

187. Ouchi, S., Oku, H., Hibino, C., and Akiyama, I. 1974. Induction of accessibility to a nonpathogen

by preliminary inoculation with a pathogen. Phytopathol. Z. 79:142-154.
188. Parlevliet, J. E. 1975. Partial resistance of barley to leaf rust, Puccinia hordei. 1. Effect of cultivar and development stage on latent period. Euphytica 24:21-27.
189. Parlevliet, J. E. 1976. Partial resistance of barley to leaf rust, Puccinia hordei. III. The inheritance of the host plant effect on latent period in four cultivars. Euphytica 25:241-248.
190. Parlevliet, J. E. 1976. Evaluation of the concept of horizontal resistance in the barley: Puccinia hordei host-pathogen relationship. Phytopathology 66:494-497.
191. Parlevliet, J. E. 1980. Minor genes for partial resistance epistatic to the Pa_7 gene for hypersensitivity in the barley: Puccinia hordei relationship. pp. 53-57 in: Proc. 5th European and Medn. Cereals Rusts Conf. Bari, Italy.
192. Parlevliet, J. E. 1983. Models explaining the specificity and durability of host resistance derived from the observations on the barley-Puccinia hordei system. pp. 57-78 in: Durable Resistance in Crops. F. Lamberti, J. M. Waller and N. A. Van der Graaff, eds. Plenum Press, New York.
193. Peberdy, J. F. 1980. Protoplast fusion - a new approach to interspecies genetic manipulation and breeding in fungi. pp. 63-72 in: Advances in Protoplast Research. L. Ferenczy and G. L. Farkas, eds. Pergamon, Oxford.
194. Porcelli-Armenise, V., Scaramuzzi, F., and De Gaetano, A. 1976. Associations in vitro de tissus cambiaux de quelques plantes appartenant à la famille des Oleaceae. C. R. Acad. Sci. Paris, Series D 282:851-854.
195. Priestley, A. L. 1983. Factors contributing to the stability of host resistance to oat mildew. Ph.D. Thesis, University of Wales, Aberystwyth.
196. Pringle, R. B., and Scheffer, R. P. 1964. Host-specific plant toxins. Annu. Rev. Phytopathol. 2:133-156.
197. Rajasekhar, E. W., Chatterjee, S., and Eapon, S.

1980. Fusion of plant protoplasts with amoeba induced by polyethylene glycol. Cytologia 45:149-155.

198. Rohringer, R., and Heitefuss, R. 1984. Histology and molecular biology of host-parasite specificity. pp. 193-223 in: The Cereal Rusts, Vol. 1. W. R. Bushnell and A. P. Roelfs, eds. Academic Press, New York.

199. Rohringer, R., Kim, W. K., and Samborski, D. J. 1979. A histological study of interactions between avirulent races of stem rust and wheat containing resistance genes Sr_5, Sr_6, Sr_8 or Sr_{22}. Can. J. Bot. 57:321-324.

200. Salmon, E. W. 1905. On endophytic adaptation shown by Erysiphe graminis D.C. under cultural conditions. Ann. bot. 19:444-446.

201. Sargent, C., and Gay, J. L. 1977. Barley epidermal apoplast structure and modification by powdery mildew contact. Physiol. Plant Pathol. 11:195-205.

202. Savile, D. B. O. 1968. The rusts of Cheloneae (Scrophulariaceae): a study in the co-evolution of hosts and parasites. Nova Hedwigia XV:369-392.

203. Scheffer, R. P. 1976. Host-specific toxins in relation to pathogenesis and disease resistance. pp. 247-269 in: Encyclopedia of Plant Physiology. R. Heitefuss and P. H. Williams, eds. Springer-Verlag, Berlin, Heidelberg and New York.

204. Scheffer, R. P., Nelson, R. R., and Ullstrup, A. J. 1967. Inheritance of toxin production and pathogenicity in Cochliobolus carbonum and Cochliobolus victoriae. Phytopathology 57:1288-1291.

205. Schmelzer, E., Borner, H., Grisebach, H., Ebel, J., and Hahlbrock, K. 1984. Phytoalexin synthesis in soybean (Glycine max). Similar time courses of mRNA induction in hypocotyls infected with a fungal pathogen and in cell cultures treated with fungal elicitor. FEBS Lett. 172:59-63.

206. Schonbeck, F., and Schlosser, E. 1976. Preformed substances as potential protectants.

pp. 653-678 in: Physiol. Plant Pathol. R. Heitefuss and P. H. Williams, eds. Springer-Verlag, Berlin.

207. Schrauwen, J. A. M. 1984. Cellular interaction in plasmodial slime moulds. pp. 291-308 in: Cellular Interactions, Encyclopedia of Plant Physiology, New Series Vol. 17. H. F. Linskens and J. Heslop-Harrison, eds. Springer-Verlag, Berlin.

208. Sequeira, L. 1979. Recognition between plant hosts and parasites. pp. 71-84 in: Host-Parasite Interfaces. B. B. Nickol, ed. Academic Press, New York.

209. Shepherd, K. W., and Mayo, G. M. E. 1972. Genes conferring specific plant disease resistance. Science 175:375-380.

210. Sidorov, V. A., Gleba, Y. Y., Krivokhat-Skaya, L. D., and Sytnik, K. M. 1978. Ultrastructural analysis of the fusion of cells of humans with protoplasts of higher plants. Akademia Nauk SSSR Doklady Biological Sci. Sect. 240;212-214.

211. Skipp, R. A., Harder, D. E., and Samborski, D. J. 1974. Electron microscopy studies on infection of resistant (Sr_6 gene) and susceptible near-isogenic wheat lines by Puccinia graminis f. sp. tritici. Can. J. Bot. 52:2615-2620.

212. Slesinski, R. S., and Ellingboe, A. H. 1969. The genetic control of primary infection of wheat by Erysiphe graminis f. sp. tritici. Phytopathology 59:1833-1837.

213. Smith, G. 1900. The haustoria of the Erysipheae. Bot. Gaz. 29:153-183.

214. Stall, R. E. 1985. Plasmid-specified host specificity in Xanthomonas campestris pv. vesicatoria. Abstr. of the VI Int. Conf. on Plant Pathogenic Bacteria, Univ. of Maryland, College Park, MD, June, 1985, p. 64.

215. Staskawicz, B. J., Dahlbeck, D., and Keen, N. T. 1984. Cloned avirulence gene of Pseudomonas syringae pv. glycinea determines race-specific incompatibility on Glycine max (L.) Merr. Proc. Natl. Acad. Sci. (U.S.A.) 81:6024-6028.

216. Stebbins, G. L., Jr. 1950. Variation and

Evolution in Plants. Columbia University, New York.
217. Strange, R. N., Majer, J. R., and Smith, H. 1974. The isolation and identification of choline and betaine as the two major components in anthers and wheat germ that stimulate Fusarium graminearum in vitro. Physiol. Plant Pathol. 4:277-290.
218. Tani, T., and Mayama, S. 1982. Evaluation of phytoalexins and preformed antifungal substances in relation to fungal infection. pp. 301-314 in: Plant Infection. Y. Asada, W. R. Bushnell, S. Ouchi, and C. P. Vance, eds. Japan Scientific Societies Press, Toyko, and Spring-Verlag, Berlin, Heidelberg, and New York.
219. Teasdale, J., Daniels, D., Davis, W. D., Eddy, R., and Hadwiger, L. E. 1974. Physiological and cytological similarities between disease resistance and cellular incompatibility responses. Plant Physiol. 54:690-695.
220. Tepper, C. S., and Anderson, A. J. 1984. Purification of fungal elicitors. Phytopathology 74:850 (Abstr.).
221. Tietjen, K. G., and Matern, U. 1984. Induction and suppression of phytoalexin biosynthesis in cultured cells of safflower, Carthamus tinctorus L., by metabolites of Alternaria carthami Cowdhury. Arch. Biochem. Biophys. 229:136-144.
222. Vance, C. P., Kirk, T. K., and Sherwood, R. T. 1980. Lignification as a mechanism of disease resistance. Annu. Rev. Phytopathol. 18:259-288.
223. VanEtten, H. D. 1982. Phytoalexin detoxification by monooxygenases and its importance for pathogenicity. pp. 315-327 in: Plant Infection: The Physiological and Biochemical Basis. Y. Asada, W. R. Bushnell, S. Ouchi, and C. P. Vance, eds. Japan Scientific Societies Press, Tokyo, and Springer-Verlag, Berlin, Heidelberg, and New York.
224. Walton, J. D., and Earle, E. D. 1984. Characterization of the host-specific phytotoxin victorin by high-pressure liquid chromatography. Plant Sci. Lett. 34:231-238.

225. Ward, M., Davey, M. R., Mathias, R. J., Cocking, E. C., Clothier, R. H., Balls, M., and Luch, J. A. 1979. Effects of pH, Ca^{2+}, temperature, and protease pretreatment on interkingdom fusion. Somatic Cell Genetics 5:529-536.
226. Whipps, J. M., Clifford, B. C., Roderick, H. W., and Lewis, D. H. 1980. A comparison of development of Puccinia hordei Otth. on normal and slow rusting varieties of barley (Hordeum vulgare L.) using analysis of fungal chitin and mannan. New Phytologist 85:191-199.
227. Willis, G. E., Hartmann, J. X., and DeLamater, E. D. 1977. Electron microscopy study of plant-animal cell fusion. Protoplasma 91:1-14.
228. Wolpert, T. J., Macko, V., Acklin, W., Juan, B., Seibl, J., Meili, J., and Arigoni, D. 1985. Structure of the host-selective toxins produced by Cochliobolus victoriae. (Abstr.) Phytopathology 75: (in press).
229. Wood, R. K. S. 1976. Specificity - an assessment. pp. 327-338 in: Specificity in Plant Disease. R. K. S. Wood and A. Graniti, eds. Plenum Press, New York and London.
230. Yang, S. L., and Ellingboe, A. H. 1972. Cuticle layer as a determining factor for the formation of mature appressoria of Erysiphe graminis on wheat and barley. Phytopathology 62:708-712.
231. Yeoman, M. M. 1984. Cellular recognition systems in grafting. pp. 453-490 in: Cellular Interactions, Encyclopedia of Plant Physiology, New Series. Vol. 17. H. F. Linskens and J. Heslop-Harrison, eds. Springer-Verlag, Berlin.
232. Yeoman, M. M., Kilpatrick, D. C., Miedzybrodzka, M. B., and Gould, A. R. 1978. Cellular interaction during graft formation in plants, a recognition phenomenon? pp 139-160 in: Cell-Cell Recognition, Symp. Soc. Exp. Biol. Vol. 32. A. S. G. Curtis, ed. Cambridge Univ., Cambridge.
233. Yoder, O. C., and Scheffer, R. P. 1969. Role of toxin in early interactions of Helminthosporium victoriae with susceptible and resistant oat tissue. Phytopathology 59:1954-1959.
234. Yoshikawa, M., Masago, H., and Keen, N. T. 1977.

Activated synthesis of poly(A)-containing messenger RNA in soybean hypocotyls inoculated with Phytophthora megasperma var. sojae. Physiol. Plant Pathol. 10:125-138.
235. Yoshikawa, M., Madama, M., and Masago, H. 1981. Release of a soluble phytoalexin-elicitor from mycelial walls of Phytophthora megasperma var. sojae by soybean tissues. Plant Physiol. 67:1032-1035.
236. Yoshikawa, M., Keen, N. T., and Wang, M-C. 1983. A receptor on soybean membranes for a fungal elicitor of phytoalexin accumulation. Plant Physiol. 73:497-506.
237. Zeyen, R. J., and Bushnell, W. R. 1979. Papilla response of barley epidermal cells caused by Erysiphe graminis: rate and method of deposition determined by microcinematography and transmission electron microscopy. Can. J. Bot. 57:898-913.
238. Zeyen, R. J., Carver, T. L. W., and Ahlstrand, G. G. 1983. Relating cytoplasmic detail of powdery mildew infection to presence of insoluble silicon by sequential use of light microscopy, S.E.M., and X-ray microanalysis. Physiol. Plant Pathol. 22:101-108.
239. Ziegler, E., and Pontzen, R. 1982. Specific inhibition of glucan-elicited glyceollin accumulation in soybeans by an extracellular mannan-glycoprotein of Phytophthora megasperma f. sp. glycinea. Physiol. Plant Pathol. 20:321-331.

INDEX